Lecture Notes in Mathematics

Edited by A. Dold and B. Eckmann

869

Andrey V. Zelevinsky

Representations of Finite Classical Groups

A Hopf Algebra Approach

Springer-Verlag
Berlin Heidelberg New York 1981

Author

Andrey V. Zelevinsky
Institute of Physics of the Earth
Department of Applied Mathematics
B. Grouzinskaya 10, 123810 Moscow, USSR

AMS Subject Classifications (1980): 16 A 24, 20 C 30, 20 G 05, 20 G 40

ISBN 3-540-10824-6 Springer-Verlag Berlin Heidelberg New York
ISBN 0-387-10824-6 Springer-Verlag New York Heidelberg Berlin

Printed in Germany

Printing and binding: Beltz Offsetdruck, Hemsbach/Bergstr.
2141/3140-543210

Contents

Introduction

In this work we develop a new unified approach to the representation theory of symmetric groups and general linear groups over finite fields. It gives an explanation of the well known non-formal statement that the symmetric group is "the general linear group over the (non-existent) one element field". This approach is based on the structural theory of a certain class of Hopf algebras.

The original plan of this work was to apply the technique developed by J.N.Bernstein and the author for the investigation of representations of general linear groups over p-adic fields ([1],[2]) to the representation theory of the groups $GL(n,F_q)$. The main tool of [1], [2] is the systematic use of the functors $i_{v,\theta}$ and $r_{v,\theta}$ and the general theorem on their composition ([1], §§ 1,5). These functors generalize those of induction and restriction while the composition theorem generalizes the Mackey theorem. The main results of [1] and [2] were obtained by the formal manipulations with these functors.

This technique may be applied to the groups $GL(n,F_q)$ without any difficulties. Moreover, the simplifications caused by the full reducibility of representations, allow one to obtain for finite groups the more complete results than for p-adic ones.

For example, in [3] there was described the restriction of representations of $GL(n,F_q)$ to the subgroup P_n consisting of matrices with the last row $(0,0,\dots,0,1)$; in p-adic case such a description is still unknown. Unfortunately, the reasonings in [3] are not self-contained: they use the relationships between the representations of $GL(n, F_q)$ and those of the symmetric groups S_n as well as the representation theory of the symmetric groups. In fact, the technique of [1], [2] can be applied directly to the representation theory of S_n. This approach allows one to obtain from the unified point of view the main classical results.

It turns out that the most convenient language in our approach to the representation theory of $GL(n, F_q)$ and S_n is that of the Hopf algebra. Consider the sequence of the groups G_n $(n \geqslant 0)$, where G_n is either $GL(n, F_q)$ (q is fixed) or S_n. Let $R(G_n)$ be the Grothendieck group of the category of finite-dimensional complex representations of G_n, and $R = \bigoplus_{n \geqslant 0} R(G_n)$. Functors of the form $i_{v,\theta}$ endow R with the structure of an algebra over $\mathbb{Z}$, while functors $r_{v,\theta}$ make R into a coalgebra (in the case of symmetric groups one has the usual functors of induction and restriction, see 1.1 below). The composition theorem transforms into the statement that R is a Hopf algebra i.e. that the comultiplication $m^*: R \longrightarrow R \otimes R$ is a ring homomorphism. The ring structure on R in the case of symmetric groups was considered in classical works by A.Young while for $GL(n, F_q)$ it was introduced by J.Green [4]. Although the Hopf algebra structure on R is also very natural, its existence for S_n was mentioned only quite recently (see [5]) while for $GL(n, F_q)$ it seems to be new.

The Hopf algebra R satisfies two extra axioms of positivity and self-adjointness. The positivity means that multiplication and comultiplication are induced by the operations on ordinary representations, i.e. they take representations to representations (not only to virtual ones). The self-adjointness means that the functor $r_{V,\theta}$ is adjoint to $i_{V,\theta}$ (a generalization of Frobenius reciprocity). The formal definitions will be given in 1.4. The crucial observation for this work is that all results on representations of the groups G_n, obtained by our method are in fact based only on these properties of the Hopf algebra R. So the problem arises to develop the structural theory of Hopf algebras satisfying the positivity and self-adjointness axioms (we call them PSH-algebras). This problem is completely solved in Chapter I of the present paper. Chapters II and III contain some applications of this theory.

Before a detailed description of the contents of this paper let us make three remarks.

1) If we put $G_n = GL(n, F)$, where F is a p-adic field, then the Hopf algebra structure on R and its positivity remain valid while the self-adjointness fails. It would be very interesting to find a weakened form of the PSH-algebra axioms such that the corresponding structural theory includes the representation theory of the groups GL(n, F). Note that many of results of the present work related to the representations of $GL(n, F_q)$, have p-adic analogues ([2], [6]).

2) This work is closely connected with that of D.K.Faddeev [7]. In [7], the representation theory of the groups

$GL(n, F_q)$ is treated by an elementary method based on the theorem on the intertwining number of two induced representations. This theorem in the case of finite groups is equivalent to our composition theorem, so the method of D.K.Faddeev is similar to our one. Some of the arguments in Chapter I are similar to those of [7] and this Chapter can be considered as an axiomatization of [7].

3) This work is entirely elementary and practically self-contained. In Chapter I besides usual linear algebra, we use only some general properties of Hopf algebras, which may be found in [8]; for convenience, these properties are given in Appendix 1. In Chapters II and III only some basic facts on the representations of finite groups are needed; the first half of the book [9] is more than sufficient.

Now we describe the contents of this paper. As indicated above, Chapter I (§§ 1-4) is devoted to the theory of PSH-algebras. In § 1 the axioms of PSH-algebras are given and some preliminary results are proved. First of all we consider the model example of representations of symmetric groups which motivates the subsequent formal definitions. The structural theory of PSH-algebras is developed in §§ 2-4. In § 2 we prove that every PSH-algebra naturally decomposes into the tensor product of "elementary" ones, namely of PSH-algebras with only one irreducible primitive element (Theorem 2.2). In § 3 we prove that an "elementary" PSH-algebra R is unique up to an isomorphism and change of grading. Furthermore the Hopf algebra structure on R is described explicitly (Theorem 3.1). This universal PSH-algebra R is studied in detail in §§ 3-4. It has some natural bases. Each of them is parametrized by partitions. We compute explicitly all transition matrices between these bases as well as the action of multiplication and comultiplication on them. Note that R can be realized as the representation algebra of the symmetric groups so all results on R can be reformulated in terms of representations of symmetric groups. It is useful to keep this in mind in the course of reading §§ 3,4, although the applications to symmetric groups are discussed separately in § 6.

Chapters II and III contain various applications of the results of Chapter I. Although the notion of a PSH-algebra is motivated by representations of the groups S_n and $GL(n, F_q)$, it has some another applications (an usual advantage of the axiomatic method!). In this work two more applications are given, to symmetric polynomials and to representations of wreath products. I am sure that another applications are possible e.g. to topology and representations of real and complex groups, and hope to consider them in the future.

In § 5 the algebra R from §§ 3,4 is naturally identified with the algebra A of symmetric polynomials in the countable set of indeterminates over $\mathbb{Z}$. This realization induces some interesting structures on A. As the immediate corollaries of results of §§ 3,4 we obtain new proofs of some classical results on symmetric polynomials e.g. the Aitken theorem, the Littlewood-Richardson rule and the Littlewood-Roe theorem. Section 5 is closely connected with the paper of L.Geissinger [5] where the Hopf algebra structure on A and the property of self-adjointness are discussed. Note that in [5] some interesting applications are given. It would be interesting to compare the present approach to the ring A with that of D.Knutson, based on the theory of λ - rings (see [10]).

In § 6 the representations of symmetric groups are considered. In 6.3. we give a dictionary translating notions of Chapter I to the language of the symmetric groups. This dictionary allows one to derive from the results of §§ 3,4 all main classical results on the representations of the symmetric groups including the Branching rule and the Murnaghan-Nakayama and Frobenius Character formulas. We conclude § 6 with the "Hook formula" for dimensions of irreducible representations of S_n (see [11]).

In § 7 we extend the results of § 6 from symmetric groups to wreath products. If G is a finite group then the wreath product $S_n[G]$ is defined to be the semidirect product of S_n with G^n, on which S_n acts via permutations of coordinates (see 7.1 below). For example, if G consists of 2 elements then $S_n[G]$ is the hyperoctahedral group, i.e. the Weyl group of type C_n (or B_n). As in the case of symmetric groups, the group $R(S[G]) = \bigoplus_{n \geq 0} R(S_n[G])$ is naturally endowed with the structure of a PSH-algebra. Its irreducible primitive elements correspond to irreducible representations of G, so for its investigation one must apply the decomposition theorem from § 2. We give the classification of irreducible representations of the groups $S_n[G]$ and compute the character table of $S_n[G]$ in terms of that of G. Note that the classification of irreducible representations of $S_n[G]$ and even more general wreath products is due to W.Specht; for detailed exposition see [12]. I hope the present approach is in some aspects more transparent.

In Chapter III we apply the theory of PSH-algebras to the representation theory of general linear and affine groups over a finite field. In § 8 we define the functors $i_{V,\theta}$ and $r_{V,\theta}$ and obtain their main properties. In § 9 the classification of irreducible representations of the groups $GL(n, F_q)$ in terms of cuspidal ones is given. Fix a finite field F_q and set $G_n = GL(n, F_q)$ $(n = 0,1,2,\ldots)$. Set $R(q) = \bigoplus_{n\geq 0} R(G_n)$; the functors of the form $i_{V,\theta}$ and $r_{V,\theta}$ endow $R(q)$ with the structure of a PSH-algebra (9.1). The results of § 2 imply that $R(q)$ decomposes into the tensor product of subalgebras $R(\rho)$, where ρ ranges over cuspidal representations of the groups G_n (9.2, 9.3). Note that the important notion of a cuspidal representation appears very naturally in our approach.

According to the results of § 3, each of algebras $R(\rho)$ is isomorphic (up to change of grading) to the algebra $R(S)$ of representations of the symmetric groups. This result was proved by D.K.Faddeev [7] by a similar method. There are two ways to identify $R(S)$ with $R(\rho)$ as PSH-algebras. To choose one of them we use the results by I.M.Gelfand and M.I.Graev [13] and S.I.Gelfand [14]. Let U_n be a maximal unipotent subgroup of G_n and ψ a nondegenerate character of

U_n. For each representation π of G_n denote by $\delta(\pi)$ the dimension of the subspace of vectors v in π such that $\pi(u)v = \psi(u)\cdot v$ for all $u \in U_n$; π is called nondegenerate if $\delta(\pi) \neq 0$ and degenerate otherwise. I.M.Gelfand and M.I.Graev proved that $\delta(\omega) \leq 1$ for each irreducible representation ω of G_n while S.I.Gelfand proved that all cuspidal representations are nondegenerate. We determine the isomorphism of $R(S)$ and $R(\rho)$ uniquely by the requirement that it takes the identity representation of S_n to a degenerate representation in $R(\rho)$. Note that in § 11 we give independent proofs of theorems of I.M.Gelfand and M.I.Graev and S.I.Gelfand. They make our approach to the representations of G_n entirely self-contained.

In § 10 the P.Hall algebra $\mathcal{H}$ is considered. By definition, $\mathcal{H} = \bigoplus_{n \geq 0} \mathcal{H}_n$, where $\mathcal{H}_n$ is the space of complex-valued class functions on G_n supported on the unipotent elements. One has the projection $p: R(q) \longrightarrow \mathcal{H}$ assigning to each representation of G_n the restriction of its character to unipotent elements. We endow $\mathcal{H}$ with the structure of a Hopf algebra over $\mathbb{C}$ such that p becomes a Hopf algebra morphism. We prove that the restriction of p to the Hopf subalgebra $R(\iota) \subset R(q)$ generated by the identity representations of the groups G_n, gives the isomorphism $R(\iota) \otimes \mathbb{C} \xrightarrow{\sim} \mathcal{H}$. (Theorem 10.3). By this isomorphism we identify $\mathcal{H}$ with $R \otimes \mathbb{C}$ where R is the universal PSH-algebra from §§ 3,4. This allows one to obtain very simply a lot of results on $\mathcal{H}$; in particular, the Green polynomials arise naturally, and we obtain their main properties.

At the beginning of § 11 the theorems of I.M.Gelfand and M.I.Graev and S.I.Gelfand are proved by means of the technique of § 10. Then we obtain the J.Green formula for the values of irreducible characters of G_n at unipotent elements. In our terms, the problem is to compute explicitly the morphism $p: R(q) \longrightarrow \mathcal{H}$; it is done in Theorem 11.7. As a corollary, we compute very simply dimensions of irreducible representations of G_n and their character values at regular unipotent elements. (Proposition 11.10); we give also a very simple proof of the Macdonald conjecture for GL(n) (Prop.11.11). It would be very interesting to obtain the complete J.Green Character Formu;a by methods of this work.

All results of §§ 10,11 are well-known (see [4] , [15]), but I hope that the present approach makes them considerably more transparent. The conclusive §§ 12 and 13 contain more fresh results.

In § 12 we consider the representations of G_n induced from various one-dimensional representations of the maximal unipotent subgroup U_n (we call them degenerate Gelfand-Graev modules). I.M.Gelfand and M.I.Graev in [13] proved that every irreducible representation of G_n can be embedded in one of these modulas. We obtain the more precise result computing decomposition of these modules into irreducible components (Theorem 12.1). As a corollary, we construct for each irreducible representation ω of G_n the degenerate Gelfand-Graev module containing ω with multiplicity 1 (Proposition 12.5); this realization of ω is an analogue of a degenerate Whittaker model for p-adic groups, obtained in [2] .

As an application of this realization, we prove that the Schur index of each irreducible representation of G_n equals 1 (Proposition 12.6); for char $F_q \neq 2$ this was proved by Z.Ohmori [16] in a considerably more complicated way.

In § 13 the relationships between the representations of the groups G_n and those of the general affine groups P_n (see above) are considered. The group P_n decomposes into the semidirect product of G_{n-1} with the abelian normal subgroup V_n. The classification of irreducible representations of P_n is easily derived from the general representation theory of such products ([9] , 9.2). Irreducible representations of P_n happen to be in a natural one-to-one correspondence with irreducible representations of all groups $G_{n-1}, G_{n-2}, \ldots, G_0$ (see 13.1, 13.2; another proof is due to D.K.Faddeev [17]). In terms of this classification, we compute explicitly the restriction of irreducible representations of G_n to P_n and of P_n to G_{n-1} (Theorem 13.5). It is interesting that these restrictions always are multiplicity-free. As a corollary, we describe explicitly the restriction of irreducible representations of G_n to G_{n-1} (Corollary 13.8); this restriction was computed by E.Thoma [18] by a quite different method and in quite different terms.

The results of § 12 seem to be new; a half of Theorem 13.5 was announced in [3] .

Some technical results are collected in 3 Appendices. In Appendix I we prove all general statements on Hopf algebras needed in this work. In Appendix 2 a combinatorial proposition is proved, on which our proof of the Littlewood-Richardson Rule is based. I believe it is of independent interest. We use the beautiful reformulation of the Littlewood-Richardson Rule given in [19] . Using it, the author recently has obtained the generalization of this rule [20] (see Remark A2,6 in the end of Appendix 2). In Appendix 3 the general theorem on the composition of functors $i_{U,\theta}$ and $r_{V,\psi}$ is stated and all its applications used in this work are collected together.

It is a pleasure to express my deep gratitude to J.N.Bernstein who played very important role in this work. He has read carefully a number of original versions of this work and suggested a lot of valuable improvements (the main ones are referred in the body of the paper). It was J.N.Bernstein who suggested to me to apply the theory of PSH-algebras to general wreath products (in the original version only hyperoctahedral groups were considered), degenerate Gelfand-Graev modules, the Schur index of representations of G_n and to the new proof of the Gelfand-Graev theorem.

At the various stages of this work it was presented at the seminars of A.M.Vershik (Leningrad State University) and of D.B.Fuchs (Moscow State University). I am grateful to A.M.Vershik, D.B.Fuchs and the participants of their seminars for their interest in this work.

Chapter I. Structural theory of PSH-algebras

§ 1. Definitions and first results

1.1. First we discuss the basic model example concerning representations of symmetric groups (in more detail it will be discussed in § 6). After classical works of A.Young it becomes clear that one must study complex representations of all these groups together, taking into account their interaction. This interaction is carried out by operations of induction and restriction.

Let S_n be the permutation group of the set $[1,n] = \{1,2,\dots,n\}$. For $k+l=n$ the group $S_k \times S_l$ is naturally embedded in S_n as the stabilizer of the subset $[1,k] \subset [1,n]$. It allows us to construct from representations π of S_k and ρ of S_l the representation of S_n induced from the tensor product representation $\pi \otimes \rho$ of the subgroup $S_k \times S_l \subset S_n$. Conversely, one may restrict a representation σ of S_n to the subgroup $S_k \times S_l$; since any irreducible representation of $S_k \times S_l$ is a tensor product of representations of S_k and S_l, one obtains a sum of such tensor products.

Let $R(S_n)$ be the Grothendieck group of the category of finite dimensional complex representations of S_n; it is a free abelian group generated by the equivalence classes of irreducible representations of S_n. The tensor product gives an isomorphism $R(S_k \times S_l) = R(S_k) \otimes R(S_l)$ and the operations of induction and restriction described above give rise to the $\mathbb{Z}$-linear maps

$$i_{k,l}: R(S_k) \otimes R(S_l) \to R(S_n) \quad \text{and} \quad r_{k,l}: R(S_n) \to R(S_k) \otimes R(S_l)$$

It is convenient to consider all these maps together. Consider the graded group $R(S) = \bigoplus_{n \geq 0} R(S_n)$ (here $S_o = \{e\}$, so $R(S_o) = \mathbb{Z}$). Define graded group morphisms

$$m : R(S) \otimes R(S) \longrightarrow R(S) \quad \text{and} \quad m^* : R(S) \longrightarrow R(S) \otimes R(S)$$

by

$$m \Big|_{R(S_k) \otimes R(S_l)} = i_{k,l}, \quad m^* \Big|_{R(S_n)} = \sum_{k+l=n} r_{k,l}$$

We consider m as a multiplication; it makes $R(S)$ into a graded algebra over $\mathbb{Z}$. Similarly, m^* is a comultiplication making $R(S)$ into a coalgebra. These structures happen to be compatible and they make $R(S)$ into a Hopf algebra over $\mathbb{Z}$ (for definition of a Hopf algebra see [8]; we shall recall it later). The fact that Hopf algebra axioms are valid is non-trivial; it expresses in a condensed form the essential properties of representations. The most essential is the statement that m and m^* are compatible, i.e. that $m^* : R(S) \to R(S) \otimes R(S)$ is a ring homomorphism. Really, to prove this one has to compute the composition $m^* \circ m$, i.e. the composition $r_{k',l'} \circ i_{k,l}$ for all k, l, k', l'. The computation is based on the Mackey Theorem on restriction of induced representations (for detail see Appendix 3).

The Hopf algebra structure on $R(S)$ accumulates essential information on representations of S_n. But this structure doesn't differ ordinary representations from virtual ones. Consider the $\mathbb{Z}$-basis Ω in $R(S)$ consisting of irreducible representations of all groups S_n, and the basis $\Omega \times \Omega$ in $R(S) \otimes R(S)$. Ordinary representations of the groups S_n ($S_k \times S_l$) correspond to positive elements in $R(S)$ ($R(S) \otimes R(S)$), i.e. to non-negative linear combinations of elements of Ω ($\Omega \times \Omega$). The important property that induction and restriction take representations to representations, means that multiplication and comultiplication are positive i.e. take positive elements to positive ones.

By Frobenius reciprocity the functors of induction and restriction are adjoint to each other. To express this fundamental fact in terms of the group $R(S)$ we consider the $\mathbb{Z}$-valued bilinear form $\langle\ ,\ \rangle$ on $R(S)$ such that all subgroups $R(S_n)$ are mutually orthogonal and for representations π, ρ of the same S_n we have

$$\langle \pi, \rho \rangle = \dim \operatorname{Hom}_{S_n}(\pi, \rho).$$

The Schur Lemma implies that Ω is an orthonormal basis of $R(S)$. Clearly, the inner product $\langle\ ,\ \rangle$ on $R(S) \otimes R(S)$ which has $\Omega \times \Omega$ as an orthonormal basis, has the same meaning in terms of representations. Therefore the Frobenius reciprocity means that operators m and m^* are adjoint to each other i.e. that $\langle x, m(y) \rangle = \langle m^*(x), y \rangle$ for $x \in R(S), y \in R(S) \otimes R(S)$.

1.2. Now we give formal definitions motivated by the preceding example. A <u>trivialized group</u> (briefly <u>T-group</u>) is a free $\mathbb{Z}$-module R with a distinguished $\mathbb{Z}$-basis $\Omega = \Omega(R)$. We consider $\mathbb{Z}$ as a T-group with $\Omega(\mathbb{Z}) = \{1\}$. The direct sum of each family of T-groups and the tensor product of a finite family of T-groups become themselves T-groups via

$$\Omega(\bigoplus_{\alpha\in A} R_\alpha) = \coprod_{\alpha\in A} \Omega(R_\alpha); \quad \Omega(\bigotimes_{i=1}^{n} R_i) = \prod_{i=1}^{n} \Omega(R_i)$$

In particular, any T-group R decomposes into the direct sum of T-groups $\mathbb{Z}\cdot\omega$, $\omega \in \Omega(R)$. A T-subgroup of a T-group R is any subgroup of the form $\bigoplus_{\omega\in\Omega'} \mathbb{Z}\cdot\omega$, where Ω' is a subset of $\Omega(R)$.

Put

$$R^+ = \left\{ \sum_{\omega\in\Omega} m_\omega \cdot \omega \;\middle|\; m_\omega \geq 0 \right\};$$

elements of R^+ are called positive. We write $x \geqslant y$ if $x - y \in R^+$. A homomorphism between two T-groups is called positive (or T-group morphism) if it takes positive elements to positive ones.

For each T-group R define the $\mathbb{Z}$-valued bilinear form $\langle\,,\rangle$ on R by

$$\langle\omega, \omega'\rangle = \delta_{\omega,\omega'} \quad \text{for } \omega, \omega' \in \Omega.$$

The form $\langle\,,\rangle$ is symmetric, nondegenerate and positively defined; we call it an inner product on R. We will freely use geometrical terminology; e.g. elements of Ω may be characterized as positive elements of length 1.

Elements of $\Omega(R)$ are called irreducible elements of a T-group R. If $\pi = \sum_{\omega\in\Omega} m_\omega \cdot \omega \in R^+$ then the elements $\omega \in \Omega$ such that $m_\omega > 0$ are called irreducible constituents of π; clearly the condition $m_\omega > 0$ can be written down as $\omega \leq \pi$ or as $\langle\omega, \pi\rangle > 0$

1.3. Now we recall the terminology on Hopf algebras.

Let K be a commutative ring with unit. A <u>Hopf algebra over K</u> is a graded K-module $R = \bigoplus_{n \geq 0} R_n$ with K-module morphisms $m : R \otimes R \to R$ (multiplication), $m^* : R \to R \otimes R$ (comultiplication), $e : K \longrightarrow R$ (unit) and $e^* : R \to K$ (counit) satisfying the following six axioms (G), (A), (U), (A^*), (U^*) and (H).

(G) (Grading). Each of m, m^*, e and e^* is a morphism of graded modules ($R \otimes R$ and K are graded by $(R \otimes R)_n = \bigoplus_{k+l=n} (R_k \otimes R_l)$ and $K = K_0$).

(A) (Associativity). The multiplication m is associative.

(U) (Unit). The element $e(1) \in R_0$ is unit of the ring R.

The axioms (A^*) and (U^*) are the associativity of comultiplication (= coassociativity) and the property of counit. In general, if (X) is a property expressing the commutativity of a certain diagram D constructed by means of morphisms m and e, we write (X^*) for the property of commutativity of the diagram obtained by reversing all arrows of D and replacing m by m^* and e by e^*. For example, the axiom (U^*) means that the diagram

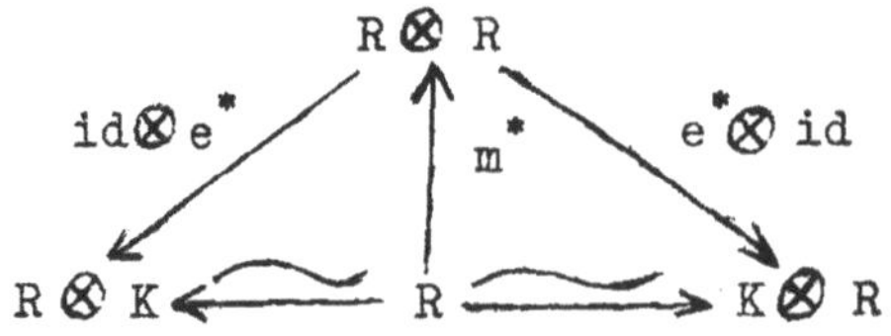

commutes.

(H) (Hopf Axiom). The operator $m^* : R \longrightarrow R \otimes R$ is a ring homomorphism (the multiplication in $R \otimes R$ is defined as usual by $(x \otimes y) \cdot (x' \otimes y') = xx' \otimes yy'$).

Removing axioms (A) and (A^*), one obtains the definition of a quasi - Hopf algebra. A (quasi) - Hopf algebra R is called <u>connected</u> if

(Con) Operators $e : K \to R_0$ and $e^* : R_0 \to K$ are mutually inverse isomorphisms.

A Hopf algebra R is called <u>commutative</u> if

(Com) The multiplication m is commutative; it is called <u>cocommutative</u> (or with commutative comultiplication) if the axiom (Com^*) holds, i.e. if the diagram

$$R \xrightarrow{m^*} R \otimes R \circlearrowleft \sigma \qquad (\sigma(x \otimes y) = y \otimes x)$$

is commutative.

1.4. Now we define the main subject of this work, a positive self-adjoint Hopf algebra (briefly PSH-algebra).

A (quasi)- Hopf algebra R over $\mathbb{Z}$ is called <u>positive</u> if it satisfies axioms (T) and (P).

(T) Each R_n and hence the whole R is a T-group (see 1.2); in other words, R is a free $\mathbb{Z}$ - module with a distinguished basis Ω consisting of homogeneous elements.

(P) (Positivity). All morphisms m, m^*, e, and e^* are positive (see 1.2).

A positive (quasi)-Hopf algebra is called <u>self-adjoint</u> if

(S) (Self-adjointness). Operators m and m^* (resp. e and e^*) are adjoint to each other with respect to inner products $\langle , \rangle$ on R, $R \otimes R$, and $\mathbb{Z}$, induced by a T-group structure (see 1.2).

A PSH-algebra is a connected positive self-adjoint Hopf algebra over $\mathbb{Z}$.

1.5. Let us introduce some notation. We denote the unit of R i.e. the element $e(1) \in R_0$ simply by 1. By axioms (Con) and (P), 1 is an irreducible element of R, and $R_0 = \mathbb{Z} \cdot 1$. We write xy for $m(x \otimes y)$ and set $I = \bigoplus_{n>0} R_n$. Axioms (G), (Con), and (U^*) imply that for $x \in I$:

$$m^*(x) = x \otimes 1 + 1 \otimes x + m_+^*(x),$$

where $m_+^*(x) \in I \otimes I$. An element $x \in I$ is called primitive if $m^*(x) = x \otimes 1 + 1 \otimes x$, i.e. $m_+^*(x) = 0$. Denote by P the subgroup of primitive elements in R. Set $I^2 = m(I \otimes I)$, i.e. I^2 is the subgroup generated by products xy, $x \in R_k$, $y \in R_l$, $k, l > 0$.

1.6. Proposition. Any positive self-adjoint quasi-Hopf algebra over $\mathbb{Z}$ is a PSH-algebra, i.e. the associativity of multiplication and comultiplication follows from other axioms of PSH-algebra. Moreover, any PSH-algebra is commutative and cocommutative.

Proof. Evidently, each of properties (A^*) and (Com^*) follows from (S) and the corresponding property "without asterisk", i.e. (A) and (Com).

Now apply the following.

1.7. Lemma. Axioms (G), (U), (U^*), (Con), (T), and (S) imply that P is the orthogonal complement of I^2 in I with respect to the inner product $\langle\, ,\rangle$.

Proof of the Lemma. According to (T), all subgroups $R_k \otimes R_l$ in $R \otimes R$ are mutually orthogonal. Hence, by (S),

$$\langle x, m(y)\rangle = \langle m^*(x), y\rangle = \langle m^*_+(x), y\rangle \quad \text{for} \quad x \in I, \quad y \in I \otimes I.$$

It follows that $x \in I$ is orthogonal to all elements of the form $m(y)$, $y \in I \otimes I$, iff $m^*_+(x) = 0$, i.e. $x \in P$.

Q.E.D.

Proposition 1.6. follows at once from this Lemma and the general theory of Hopf algebras (see Appendix 1, Lemma A1.3).

1.8. Remark. Proposition 1.6 has some applications. For example, the commutativity of the algebra $R(S)$ (see 1.1) is an essential (although rather simple) property of representations of the symmetric groups. Another less trivial application concerning representations of the groups $GL(n, F_q)$ will be given later in § 9 (see 9.1).

1.9. Let R be a PSH-algebra. For any $x \in R$ denote by $x^* : R \longrightarrow R$ the operator adjoint to that of multiplication by x i.e. x^* is defined by

$$\langle x^*(y), z\rangle = \langle y, xz\rangle \quad , \quad y, z \in R .$$

(according to Part (b) of the next proposition, x^* is well-defined). The operators x^* will be the main tool in our study of PSH-algebras. Let us summarize their main properties.

Proposition. (a) Let $x \in R_k$. Then $x^*(R_n) \subset R_{n-k}$ for $n \geq 0$; in particular, $x^*(R_n) = 0$ for $n < k$. Identifying naturally R_o with $\mathbb{Z}$ (see 1.3 (Con)) we see that the linear form x^* : $R_k \longrightarrow R_o = \mathbb{Z}$ is an inner product with x (denote it by $\langle x|$).

(b) The operator $x^* : R \longrightarrow R$ equals the composition

$$R \xrightarrow{m^*} R \otimes R \xrightarrow{id \otimes \langle x|} R \otimes \mathbb{Z} \xrightarrow{\sim} R$$

(c) For any $x, y \in R$

$$(xy)^* = y^* \circ x^* .$$

In particular, since R is commutative, all operators of the form x^* commute with each other.

(d) If $x \in R^+$ then the operator x^* is positive (see 1.2).

(e) If $x, y, z \in R$, and $m^*(x) = \sum_i a_i \otimes b_i$ then

$$x^*(yz) = \sum_i a_i^*(y)\ b_i^*(z)$$

(f) If $\rho \in R$ is primitive (see 1.5) then $\rho^* : R \longrightarrow R$ is a derivation of the ring R, i.e.

$$\rho^*(yz) = \rho^*(y) \cdot z + y \cdot \rho^*(z)$$

(g) If $\rho \in R_n$ is primitive, $0 < k < n$, and $x \in R_k$ then $x^*(\rho) = 0$.

Proof. Parts (a)-(d) follow at once from definitions while (f) is a particular case of (e). The statement (g) follows from definition of x^* and Lemma 1.7. It remains to prove (e).

Clearly, the operator $R \otimes R \longrightarrow R \otimes R$ adjoint to the multiplication by $a \otimes b$, is $a^* \otimes b^*$. Using this fact and axioms (S) and (H), we have $(u \in R)$:

$$\begin{aligned}
&\langle x^*(yz), u\rangle = \langle yz, xu\rangle = \langle m(y \otimes z), xu\rangle = \langle y \otimes z, m^*(xu)\rangle = \\
&= \langle y \otimes z,\ m^*(x) \cdot m^*(u)\rangle = \sum_i \langle y \otimes z,\ (a_i \otimes b_i) \cdot m^*(u)\rangle = \\
&= \sum_i \langle a_i^*(y) \otimes b_i^*(z),\ m^*(u)\rangle = \sum_i \langle m\left[a_i^*(y) \otimes b_i^*(z)\right],\ u\rangle = \\
&= \langle \sum_i a_i^*(y) \cdot b_i^*(z),\ u\rangle .
\end{aligned}$$

Since $u \in R$ can be chosen arbitrarily, Part (e) follows.

Q.E.D.

§ 2. The decomposition theorem

2.1. In this chapter we shall develop a complete structural theory of PSH-algebras. We shall see that this theory is quite similar to the well-known theory of Hopf algebras with commutative multiplication and comultiplication over a field K of characteristic 0. Recall that **any** such Hopf algebra $\mathcal{H}$ is naturally isomorphic to the symmetric algebra of the space of its primitive elements (see [8], § 5 or Appendix 1 below). In other words:

(1) The Hopf algebra $\mathcal{H}$ decomposes into the tensor product

$$\mathcal{H} = \bigotimes_{\alpha \in A} \mathcal{H}_\alpha$$

where each Hopf algebra $\mathcal{H}_\alpha$ has only one primitive element.

(2) The Hopf algebra $\mathcal{H}$ with only one primitive element is isomorphic to the algebra $K[x]$ of polynomials in one indeterminate x, where $\deg x = k$ and x is primitive, i.e. $m^*(x) = x \otimes 1 + 1 \otimes x$. Thus, $\mathcal{H}$ is essentially unique.

In this section we shall prove the analogue of (1) for PSH-algebras. The role of "elementary" algebras $\mathcal{H}_\alpha$ will be played by PSH-algebras with only one irreducible primitive element.

2.2. Let $(R_\alpha \mid \alpha \in A)$ be a family of PSH-algebras. Define the tensor product

$$R = \bigotimes_{\alpha \in A} R_\alpha$$

to be an inductive limit of the finite tensor products $\bigotimes_{\alpha \in S} R_\alpha$

(S ranges over finite subsets of A). Clearly, R is a PSH-algebra with the set of irreducible elements

$$\Omega(R) = \coprod_S \left(\prod_{\alpha\in S} \Omega(R_\alpha)\right);$$

each R_α is naturally embedded into R. Our first main result on PSH-algebras is the following.

Decomposition Theorem. Any PSH-algebra R decomposes into the tensor product of PSH-algebras with only one irreducible primitive element. More precisely, let $\mathcal{C} = \Omega \cap P$ be the set of irreducible primitive elements in R. For any $\rho \in \mathcal{C}$ we set

$$\Omega(\rho) = \{\omega \in \Omega \mid \langle \omega, \rho^n \rangle \neq 0 \text{ for some } n \geq 0\} \text{ and}$$

$$R(\rho) = \bigoplus_{\omega\in\Omega(\rho)} \mathbb{Z}\cdot\omega$$

Then $R(\rho)$ is a PSH-subalgebra in R with the set of irreducible elements $\Omega(\rho)$, ρ is the unique irreducible primitive element of $R(\rho)$, and R as a PSH-algebra is a tensor product $\bigotimes_{\rho\in\mathcal{C}} R(\rho)$.

This is proved in 2.3-2.7.

2.3. Proposition. Let $\rho_1, \ldots, \rho_r, \rho_1', \ldots, \rho_s' \in \mathcal{C}$ (see 2.2), $\pi = \rho_1 \cdot \ldots \cdot \rho_r$, $\pi' = \rho_1' \cdot \ldots \cdot \rho_s'$. Then

$$\langle \pi, \pi' \rangle = 0$$

unless r = s and sequences $(\rho_1, \ldots, \rho_r)$ and $(\rho_1', \ldots, \rho_s')$ are equal up to a permutation.

Proof. Apply 1.9 (f):

$$\langle \pi, \pi' \rangle = \langle \rho_2 \rho_3 \cdot \ldots \cdot \rho_r, \rho_1^*(\pi') \rangle = \langle \rho_2 \cdot \ldots \cdot \rho_r, \sum_{j=1}^{s} \rho_1' \cdot \ldots \cdot \rho_{j-1}' \cdot \rho_1^*(\rho_j') \cdot \rho_{j+1}' \cdot \ldots \cdot \rho_s' \rangle$$

By 1.9 (a), (g), if $\rho_1 \neq \rho'_j$ for all j then $\langle \pi, \pi' \rangle = 0$. Furthermore, if some of ρ'_j, say ρ'_1, equals ρ_1, and exactly k of ρ'_j equal ρ_1 then

$$\langle \pi, \pi' \rangle = k \cdot \langle \rho_1, \rho_1 \rangle \cdot \langle \rho_2 \cdot \ldots \cdot \rho_r, \rho'_2 \cdot \ldots \cdot \rho'_s \rangle .$$

The proof is completed by induction on r.

Q.E.D.

2.4. Remark. Our proof of Proposition 2.3. doesn't use the irreducibility of the elements ρ_i and ρ'_j and is based only on the fact that distinct elements among them are primitive and mutually orthogonal. This will be used later.

2.5. Denote by $S(\mathscr{C}; \mathbb{Z}^+)$ the semigroup of functions $\varphi : \mathscr{C} \to \mathbb{Z}^+$ with finite support. For any $\varphi \in S(\mathscr{C}; \mathbb{Z}^+)$ set

$$\pi_\varphi = \prod_{\rho \in \mathscr{C}} \rho^{\varphi(\rho)} \in R .$$

The axiom (P) implies that all π_φ are in R^+ (see 1.2, 1.4). Denote by $\Omega(\varphi)$ the set of irreducible constituents of π_φ and put

$$R(\varphi) = \oplus \mathbb{Z} \cdot \omega , \quad \omega \in \Omega(\varphi)$$

Proposition. (a) The set $\Omega = \Omega(R)$ is a disjoint union

$$\coprod \Omega(\varphi) , \quad \varphi \in S(\mathscr{C}; \mathbb{Z}^+) .$$

(b) The group R is graded by the semigroup $S(\mathscr{C}; \mathbb{Z}^+)$ i.e. $R = \oplus R(\varphi)$, $\varphi \in S(\mathscr{C}; \mathbb{Z}^+)$. This grading is compatible with the Hopf algebra structure, i.e.

$$m(R(\varphi') \otimes R(\varphi'')) \subset R(\varphi' + \varphi''),$$
$$m^*(R(\varphi)) \subset \bigoplus_{\varphi' + \varphi'' = \varphi} (R(\varphi') \otimes R(\varphi''))$$

Proof. (a) The disjointness of the $\Omega(\varphi)$ follows at once from 2.3. We must verify that for any $\omega \in \Omega$ there exists $\pi_\varphi \geq \omega$ (see 1.2). The statement is trivial if either $\omega \in \mathcal{C}$ or $\omega = 1$. So we suppose that $\omega \in I$ (see 1.5) and ω is not primitive. Lemma 1.7 implies that there exist two elements $\omega', \omega'' \in \Omega$ of positive degree such that $\omega \leq \omega' \cdot \omega''$.
Using induction on deg ω , one may assume that there exist $\pi_{\varphi'} \geq \omega'$ and $\pi_{\varphi''} \geq \omega''$. Hence

$$\omega \leq \omega' \cdot \omega'' \leq \pi_{\varphi'} \cdot \pi_{\varphi''} = \pi_{\varphi' + \varphi''}$$

Q.E.D.

(b) The equality $R = \bigoplus R(\varphi)$ follows at once from (a) while the inclusion

$$m(R(\varphi') \otimes R(\varphi'')) \subset R(\varphi' + \varphi'')$$

follows from the proof of (a). The corresponding property of comultiplication follows from the self-adjointness of R.

Q.E.D.

2.6. Proposition. Let $\varphi, \varphi' \in S(\mathcal{C}; \mathbb{Z}^+)$ have disjoint supports. Then multiplication

$$m : R(\varphi) \otimes R(\varphi') \longrightarrow R(\varphi + \varphi')$$

establishes an isomorphism of T-groups. In other words, the elements $\omega \cdot \omega'$ where $\omega \in \Omega(\varphi)$, $\omega' \in \Omega(\varphi')$, are irreducible, mutually distinct, and any irreducible element in $\Omega(\varphi+\varphi')$ has such a form.

Proof. Let $\omega_1, \omega_2 \in \Omega(\varphi)$, $\omega_1', \omega_2' \in \Omega(\varphi')$. We must prove that $\omega_1 \cdot \omega_1'$ is irreducible and that $\omega_1 \cdot \omega_1' \neq \omega_2 \cdot \omega_2'$ unless $\omega_1 = \omega_2$, $\omega_1' = \omega_2'$. It suffices to verify the equality

$$\langle \omega_1 \omega_1', \omega_2 \omega_2' \rangle = \delta_{\omega_1 \omega_2} \cdot \delta_{\omega_1' \omega_2'} .$$

Apply axioms (S) and (H):

$$\langle \omega_1 \omega_1', \omega_2 \omega_2' \rangle = \langle \omega_1 \omega_1', m(\omega_2 \otimes \omega_2') \rangle =$$
$$= \langle m^*(\omega_1 \omega_1'), \omega_2 \otimes \omega_2' \rangle = \langle m^*(\omega_1) \cdot m^*(\omega_1'), \omega_2 \otimes \omega_2' \rangle$$

According to 2.5 (b), $m^*(\omega_1)$ ($m^*(\omega_1')$) is a sum of components belonging to $R(\varphi_0) \otimes R(\varphi_1)$ ($R(\varphi_0') \otimes R(\varphi_1')$), where $\varphi_0 + \varphi_1 = \varphi$ ($\varphi_0' + \varphi_1' = \varphi'$). The product of such components lies in $R(\varphi_0 + \varphi_0') \otimes R(\varphi_1 + \varphi_1')$ (see 2.5 (b)). Therefore, it is orthogonal to $\omega_2 \otimes \omega_2' \in R(\varphi) \otimes R(\varphi')$ unless $\varphi_0 + \varphi_0' = \varphi$, $\varphi_1 + \varphi_1' = \varphi'$. Since $\operatorname{supp} \varphi \cap \operatorname{supp} \varphi' = \emptyset$, the equations

$$\begin{cases} \varphi_0 + \varphi_1 = \varphi, & \varphi_0' + \varphi_1' = \varphi', \\ \varphi_0 + \varphi_0' = \varphi, & \varphi_1 + \varphi_1' = \varphi' \end{cases}$$

have the unique solution

$$\varphi_0' = \varphi_1 = 0, \quad \varphi_0 = \varphi, \quad \varphi_1' = \varphi'.$$

The corresponding components of $m^*(\omega_1)$ and $m^*(\omega_1')$ equal respectively $\omega_1 \otimes 1$ and $1 \otimes \omega_1'$. Hence

$$\langle \omega_1 \omega_1', \omega_2 \omega_2' \rangle = \langle \omega_1 \otimes \omega_1', \omega_2 \otimes \omega_2' \rangle =$$
$$= \langle \omega_1, \omega_2 \rangle \cdot \langle \omega_1', \omega_2' \rangle = \delta_{\omega_1 \omega_2} \cdot \delta_{\omega_1' \omega_2'}$$

as desired.

It remains to prove that any element of $\Omega(\varphi+\varphi')$ is of the form $\omega\omega'$ where $\omega\in\Omega(\varphi)$, $\omega'\in\Omega(\varphi')$. Let

$$\pi_\varphi=\sum\omega_i \quad , \quad \pi_{\varphi'}=\sum\omega'_j$$

be the decompositions of π_φ and $\pi_{\varphi'}$ into sums of irreducible elements. Then

$$\pi_{\varphi+\varphi'}=\sum\omega_i\cdot\omega'_j$$

is the appropriate decomposition of $\pi_{\varphi+\varphi'}$, so the elements $\omega_i\omega'_j$ exhaust all irreducible constituents of $\pi_{\varphi+\varphi'}$.

Q.E.D.

2.7. Proof of Theorem 2.2. Under the notation of 2.5, the subspace $R(\rho)\subset R$ (see 2.2) equals

$$R(\rho)=\bigoplus_{n\geq 0}R(n\cdot\chi_\rho),$$

where $\chi_\rho\in S(\mathcal{C};\mathbb{Z}^+)$ is the characteristic function of the subset $\{\rho\}\subset\mathcal{C}$. By 2.5 (b) $R(\rho)$ is a PSH-subalgebra of R.

Clearly, any $\varphi\in S(\mathcal{C};\mathbb{Z}^+)$ has the form

$$\varphi=\sum_{\rho\in\mathcal{C}}n_\rho\cdot\chi_\rho \quad , \quad n_\rho\in\mathbb{Z}^+.$$

Applying several times Proposition 2.6, we see that the multiplication establishes an isomorphism of T-groups

$$m:\bigotimes_{\rho\in\mathcal{C}}R(n_\rho\cdot\chi_\rho)\xrightarrow{\sim}R(\varphi).$$

Since $R=\bigoplus R(\varphi)$, $\varphi\in S(\mathcal{C};\mathbb{Z}^+)$ (see 2.5 (b)), it follows that m establishes an isomorphism of PSH-algebras

$$\bigotimes_{\rho\in\mathcal{C}}R(\rho)\xrightarrow{\sim}R.$$

The last statement to be proved is that ρ is the unique irreducible primitive element in $R(\rho)$. This follows at once from Proposition 2.3.

Q.E.D.

§ 3. Universal PSH-algebra: the uniqueness theorem and the Hopf algebra structure

3.1. In this section we prove for PSH-algebras an analogue of the statement (2) from 2.1. We show that a PSH-algebra with only one irreducible primitive element is essentially unique, and study in detail its Hopf algebra structure.

Fix to the end of Chapter 1 a PSH-algebra R with the unique irreducible primitive element ρ. According to Theorem 2.2, any irreducible element $\omega \in R$ is an irreducible constituent of ρ^n for some $n \in \mathbb{Z}^+$. In particular, if $R_n \neq 0$ then n is divisible by $\deg \rho$. Therefore, we can change the grading on R, dividing all degrees by $\deg \rho$, and assume without loss of generality that $\deg \rho = 1$. Thus,

$$R_1 = \mathbb{Z} \cdot \rho \, .$$

Our main results on R are summarized in the following.

Theorem. (a) The element ρ^2 is a sum of two distinct irreducible elements x_2 and y_2.

(b) For any $n \geq 0$ there exist unique irreducible elements x_n and y_n in R_n such that

$$x_2^*(y_n) = 0, \quad y_2^*(x_n) = 0 \qquad \text{(see 1.9)}.$$

(c) If $0 \leq k \leq n$ then

$$x_k^* (x_n) = x_{n-k} , \quad y_k^* (y_n) = y_{n-k} .$$

If $\omega \in \Omega$ is distinct from $x_o, x_1, \ldots, x_n$ then $\omega^*(x_n) = 0$; the similar holds for y_n. In particular,

$$y_k^* (x_n) = x_k^* (y_n) = 0 \quad \text{for } k \geq 2.$$

(d) For $n \geq 1$ we have

$$m^* (x_n) = \sum_{k=0}^{n} x_k \otimes x_{n-k} ,$$

$$m^* (y_n) = \sum_{k=0}^{n} y_k \otimes y_{n-k}$$

(e) The ring R is a ring $\mathbb{Z}[x_1, x_2, \ldots]$ of polynomials in indeterminates $(x_n \mid n \geq 1)$; similarly, $R = \mathbb{Z}[y_1, y_2 \ldots]$. The elements (x_n) and (y_n) satisfy the relations

$$\sum_{k=0}^{n} (-1)^k x_k y_{n-k} = 0 \qquad (n \geq 1)$$

(f) The algebra R has a unique non-trivial PSH-algebra automorphism t. One has

$$t(x_n) = y_n \quad \text{and} \quad t(y_n) = x_n \quad \text{for } n \geq 1.$$

(g) Any PSH-algebra R' with the unique irreducible primitive element ρ' of degree 1, is isomorphic to R as a PSH-algebra; according to (f), there exist exactly two PSH-algebra isomorphisms between R and R'.

Theorem is proved in 3.2-3.14.

3.2. Proof of 3.1 (a). Apply 1.9 (a), (f):

$$\langle \rho^2, \rho^2 \rangle = \langle \rho^*(\rho^2), \rho \rangle = \langle 2\rho, \rho \rangle = 2 .$$

On the other hand, by positivity of multiplication (see 1.4) $\rho^2 \in R^+$, i.e.

$$\rho^2 = \sum_{\omega \in \Omega} m_\omega \cdot \omega \quad , \text{ where } m_\omega \in \mathbb{Z}^+ .$$

Hence

$$2 = \langle \rho^2, \rho^2 \rangle = \sum_{\omega \in \Omega} m_\omega^2 .$$

This is possible only if exactly two of m_ω are non-zero and they are equal to 1.

Q.E.D.

Note that x_2 and y_2 satisfy

$$m_+^*(x_2) = m_+^*(y_2) = \rho \otimes \rho \qquad \text{(see 1.5).}$$

Indeed, $\langle m_+^*(x_2), \rho \otimes \rho \rangle = \langle x_2, \rho^2 \rangle = 1$ (similarly for y_2).

3.3. Proof of 3.1 (b). We shall prove by induction on n the stronger statement namely that the desired x_n and y_n exist, are unique and satisfy

$$\rho^*(x_n) = x_{n-1} , \quad \rho^*(y_n) = y_{n-1} \qquad (n \geqslant 1).$$

(the latter statement is a special case of 3.1 (c)). Fix $k \geqslant 3$ and assume that the elements x_n and y_n with the desired properties are already constructed for $n < k$. We shall proceed in several steps.

(1) Let us prove that x_k must be an irreducible constituent of $\rho \cdot x_{k-1}$. Indeed, by 1.9 (d)

$$\rho^*(x_k) \in R^+ ,$$

and by 1.9 (c)

$$y_2^*(\rho^*(x_k)) = \rho^*(y_2^*(x_k)) = 0$$

Hence the uniqueness of x_{k-1} implies that

$$\rho^*(x_k) = c\cdot x_{k-1} ,$$

where $c > 0$. Therefore,

$$< x_k, \rho x_{k-1} > = < \rho^*(x_k), x_{k-1} > = c > 0,$$

as desired.

(2) Now we prove that $\rho\cdot x_{k-1}$ is a sum of two distinct irreducible elements. As in 3.2, it suffices to verify that $< \rho x_{k-1}, \rho\cdot x_{k-1} > = 2$.

Apply 1.9 (f) and the inductive assumption:

$$< \rho\cdot x_{k-1}, \rho\cdot x_{k-1} > = < \rho^*(\rho x_{k-1}), x_{k-1} > = < x_{k-1} + \rho x_{k-2}, x_{k-1} > = 1 + < \rho x_{k-2}, x_{k-1} > = 1 + < x_{k-2}, \rho^*(x_{k-1}) > =$$

$$= 1 + < x_{k-2}, x_{k-2} > = 2,$$

as desired.

(3) Now we compute $y_2^*(\rho x_{k-1})$. Apply the equality $m_+^*(y_2) = \rho\otimes\rho$, and 1.9 (e):

$$y_2^*(\rho x_{k-1}) = y_2^*(\rho)\cdot x_{k-1} + \rho^*(\rho)\cdot \rho^*(x_{k-1}) + y_2^*(x_{k-1})$$

The first and third summands are 0 by 1.9 (a) and definition of x_{k-1} . Using the inductive assumption, one obtains that

$$y_2^*(\rho x_{k-1}) = x_{k-2} .$$

(4) Since the operator y_2^* is positive (see 1.9 (d)) and takes $\rho \cdot x_{k-1}$ to an irreducible element, one obtains (with the account of (2)) that y_2^* takes one irreducible constituent of $\rho \cdot x_{k-1}$ to 0 while another one to x_{k-2}. According to (1) this proves the existence and uniqueness of x_k. The equality $\rho^*(x_k) = x_{k-1}$ follows at once from (1) and (2).

For y_n the proof is quite similar

Q.E.D.

3.4. Proof of 3.1 (c). Let $0 \leq k \leq n$. In 3.3 we proved that $\rho^*(x_n) = x_{n-1}$. Hence by 1.9 (c):

$$(\rho^k)^*(x_n) = (\rho^*)^k(x_n) = x_{n-k}.$$

According to 1.9 (d), this implies that there exists a unique irreducible $\sigma \in R_k$ such that $\sigma^*(x_n) \neq 0$; moreover, $\sigma^*(x_n) = x_{n-k}$. We must prove that $\sigma = x_k$, i.e. that $x_k^*(x_n) \neq 0$. One has

$$(\rho^{n-k})^*(x_k^*(x_n)) = x_k^*((\rho^{n-k})^*(x_n)) = x_k^*(x_k) = 1,$$

hence $x_k^*(x_n) \neq 0$, as desired.

The last assertion in 3.1 (c) follows from the fact that $x_k \neq y_k$ for $k \geq 2$. But this is obvious since

$$y_2^*(x_k) = 0 \text{ while for } k \geq 2 \quad y_2^*(y_k) = y_{k-2} \neq 0.$$

3.5. Proof of 3.1 (d). For all irreducible $\omega, \sigma \in R$ we have

$$\langle m^*(x_n), \omega \otimes \sigma \rangle = \langle x_n, \omega\sigma \rangle = \langle \omega^*(x_n), \sigma \rangle$$

(similarly for y_n). Therefore our assertion follows from 3.1(c).

3.6. Before proving the remaining assertions of Theorem 3.1, we derive some consequences from the already proven ones.

It is convenient to define x_k and y_k for all $k \in \mathbb{Z}$ by

$$x_k = y_k = 0 \quad \text{for} \quad k < 0$$

(we recall that $x_o = y_o = 1$, $x_1 = y_1 = \rho$, and $x_k \neq y_k$ for $k \geq 2$). According to 3.1 (d) and 1.9 (e), one has

$$(*) \quad \begin{cases} x_n^* (ab) = \sum\limits_{k+l=n} x_k^* (a)\, x_l^* (b), \\ y_n^* (ab) = \sum\limits_{k+l=n} y_k^* (a)\, y_l^* (b) \end{cases}$$

$(a, b \in R)$.

Define the operators X^* and Y^* on R by

$$X^* = \sum_k x_k^* \quad , \quad Y^* = \sum_k y_k^* \quad ;$$

clearly, they are well-defined and map R_n into $\bigoplus\limits_{0 \leq k \leq n} R_k$. According to $(*)$, X^* and Y^* are ring homomorphisms. They will play a crucial role in the sequel.

We define the linear forms δ_x and δ_y from R to $\mathbb{Z}$, setting for $a \in R_n$

$$\delta_x(a) = x_n^* (a), \qquad \delta_y (a) = y_n^* (a)..$$

Clearly, δ_x and δ_y are positive; by $(*)$ they are multiplicative, i.e. are ring homomorphisms. Furthermore, δ_x and δ_y satisfy the "normalization condition"

$$\delta_x(\rho) = \delta_y(\rho) = 1 .$$

The following proposition which will be useful in applications, shows that δ_x and δ_y may be characterized by these properties.

Proposition. Let $\delta : R \longrightarrow \mathbb{Z}$ be a positive, multiplicative and normalized form (i.e. $\delta(\rho) = 1$). Then δ equals either δ_x or δ_y.

Proof. Since δ is multiplicative and normalized, one has $\delta(\rho^2) = 1$. Therefore positivity of δ implies that either $\delta(x_2) = 0$ or $\delta(y_2) = 0$. Suppose that $\delta(y_2) = 0$. If $\omega \in \Omega_n, \omega \neq x_n$ then $y_2^*(\omega) \neq 0$ hence $\omega \leq y_2 \cdot \rho^{n-2}$. Therefore,

$$0 \leq \delta(\omega) \leq \delta(y_2 \cdot \rho^{n-2}) = \delta(y_2) \cdot \delta(\rho^{n-2}) = 0$$

i.e. $\delta(\omega) = 0$.

Since $\delta(\rho^n) = 1$, and $\delta(\omega) = 0$ for all irreducible constituents of ρ^n except x_n, one has $\delta(x_n) = 1$. We see that $\delta = \delta_x$. Similarly, if $\delta(x_2) = 0$ then $\delta = \delta_y$.

Q.E.D.

3.7. Now we introduce the notation and terminology relative to partitions and Young diagrams. Denote by $\mathcal{P}$ the set of families $(l_1, \ldots, l_r)$ where $l_i \in \mathbb{Z}^+$; two familles, differing by an order or the number of zeros, are identified, i.e. determine the same element of $\mathcal{P}$. Elements of $\mathcal{P}$ are called <u>partitions</u>. They will be denoted by Greek letters λ, μ, ν etc. while their parts by appropriate Latin letters e.g. $\lambda = (l_1, \ldots l_r)$. For $k \geq 1$ denote by $r_k = r_k(\lambda)$ the number of parts of λ which are equal to k; we shall sometimes write λ as $(1^{r_1}, 2^{r_2}, \ldots)$. Put

$$r(\lambda) = \sum_{k \geq 1} r_k(\lambda),$$

i.e. $r(\lambda)$ is the number of non-zero parts of λ. For each

$\lambda = (l_1, \dots, l_r)$ set

$$|\lambda| = l_1 + \dots + l_r ;$$

for $n \geqslant 0$ write $\mathcal{P}_n$ for $\{\lambda \in \mathcal{P} \mid |\lambda| = n\}$.

Any $\lambda \in \mathcal{P}$ can be written as $(l_1, \dots, l_r)$ where $l_1 \geqslant l_2 \geqslant \dots \geqslant l_r$; put also $l_s = 0$ for $s > r$. The sequence $(l_1, l_2, \dots)$ as well as each its initial segment $(l_1, l_2, \dots, l_N)$ with $N \geqslant r(\lambda)$ will be called <u>a canonical form</u> of λ (the notation is $c.f(\lambda) = (l_1, l_2, \dots)$).

A <u>Young diagram</u> is a finite subset of $\mathbb{N} \times \mathbb{N}$ containing with each point $(i.j)$ all points (i', j') such that $i' \leqslant i$, $j' \leqslant j$ (recall that $\mathbb{N}$ as usual stands for the set of positive integers). Assign to a partition $\lambda \in \mathcal{P}$ with $c.f.(\lambda) = (l_1, l_2, \dots)$ the Young diagram

$$\{(i, j) \in \mathbb{N} \times \mathbb{N} \mid j \leqslant l_i\} .$$

Clearly we obtain the bijection between $\mathcal{P}$ and the set of all Young diagrams. We identify this set with $\mathcal{P}$ via this bijection, and use the same notation for a partition and the corresponding Young diagram. For example, we write $\emptyset$ for the partition $(0) \in \mathcal{P}$.

When we show diagrams graphically, we assume the i-axis to go downwards while the j-axis to the right (as if (i, j) were a matrix index). We shall use the corresponding geometrical terminology.

The transposition $t : \mathbb{N} \times \mathbb{N} \longrightarrow \mathbb{N} \times \mathbb{N}$ acts by $(i, j)^t = (j, i)$. Evidently, t takes Young diagram to Young diagrams i.e acts on $\mathcal{P}$. Furthermore, $t^2 = id$ i.e. $(\lambda^t)^t = \lambda$ for all $\lambda \in \mathcal{P}$,

If $\lambda = (l_1,\ldots,l_r) \in \mathcal{P}$ and $l_i \neq 0$ for $i = 1,\ldots,r$ then we define the partition $\lambda^{\leftarrow} \in \mathcal{P}$ by

$$\lambda^{\leftarrow} = (l_1-1,\ l_2-1,\ldots,\ l_r-1);$$

put also $\emptyset^{\leftarrow} = \emptyset$. In terms of Young diagrams, $\lambda^{\leftarrow}$ is obtained from λ by removing the first column (and shifting by 1 to the left). Set also

$$\lambda^{\uparrow} = ((\lambda^t)^{\leftarrow})^t ;$$

this means that $\lambda^{\uparrow}$ is obtained from λ by removing the first row and shifting by 1 upwards. The following two assertions follow at once from definitions.

(1) c.f. $(\lambda^t) = (r(\lambda)\ ,\ r(\lambda^{\leftarrow}),\ r((\lambda^{\leftarrow})^{\leftarrow}),\ldots)$

(2) If c.f. $(\lambda) = (l_1,l_2,l_3,\ldots)$ then c.f.$(\lambda^{\uparrow}) = (l_2,l_3\ldots)$.

Example. $\lambda = (1,4,2,2) = (1,2^2,4)$; c.f.$(\lambda) = (4,2,2,1,0,0\ldots)$

$$\lambda = \begin{matrix} \text{xxxx} \\ \text{xx} \\ \text{xx} \\ \text{x} \end{matrix}$$

$|\lambda| = 9$; $r(\lambda) = 4$; $\lambda^t = (4,3,1,1)$; $\lambda^{\leftarrow} = (3,1,1)$; $\lambda^{\uparrow} = (2,2,1)$.

3.8. Let us return to the proof of Theorem 3.1. For any $\lambda = (l_1,\ldots,\ l_r)$ set

$$x_\lambda = x_{l_1}\, x_{l_2} \cdots x_{l_r}\ ,\quad y_\lambda = y_{l_1} \cdots y_{l_r}\ ;$$

in particular, $x_\emptyset = y_\emptyset = 1$. The first assertion in 3.1(e) means that for any $n \geqslant 0$ the monomials $(x_\lambda \mid \lambda \in \mathcal{P}_n)$ form a basis of R_n. We apply the following general

Lemma. Let $e_1,\ldots,e_p$ be elements of an arbitrary T-group R.

Then $e_1, \dots, e_p$ is a basis of a certain T-subgroup in R (see 1.2) if and only if the Gram determinant

$$\det(\langle e_i, e_j \rangle)_{i,j=1,\dots,p}$$

equals 1.

Proof. Suppose that $e_1, \dots, e_p$ form a basis of a T-subgroup $R' \subset R$. This means that $e_1, \dots, e_p$ are linearly independent and

$$R' = \mathbb{Z}\cdot\omega_1 \oplus \mathbb{Z}\cdot\omega_2 \oplus \dots \oplus \mathbb{Z}\cdot\omega_p$$

where $\omega_1, \dots, \omega_p$ are distinct irreducible elements of R. We see that the transition matrix A between the bases $\{\omega_1, \dots, \omega_p\}$ and $\{e_1, \dots, e_p\}$ is invertible, and A and A^{-1} are both integral. Therefore $\det A = \pm 1$ hence

$$\det(\langle e_i, e_j \rangle) = \det(A \cdot A^t) = (\det A)^2 = 1.$$

Conversely, let $\det(\langle e_i, e_j \rangle) = 1$. Clearly, $e_1, \dots, e_p$ are linearly independent. Let R' be the subgroup generated by $e_1, \dots, e_p$, and $R^{\perp}$ the orthogonal complement of R' in R. For any $x \in R$ consider the system of linear equations

$$\langle \sum_{i=1}^{p} a_i e_i, e_j \rangle = \langle x, e_j \rangle \quad , \; j = 1,2,\dots,p,$$

with unknowns $a_1, \dots, a_p$. The determinant of this system equals

$$\det(\langle e_i, e_j \rangle) = 1$$

so there exists the unique solution (a_i), and all a_i are integers. This means that R decomposes into the direct sum

$$R = R' \oplus R^{\perp}.$$

Let now $\omega \in R$ be an irreducible element. One has $\omega = \omega' + \omega^{\perp}$

where $\omega' \in R'$, $\omega^{\perp} \in R^{\perp}$. Therefore

$$1 = \langle \omega, \omega \rangle = \langle \omega', \omega' \rangle + \langle \omega^{\perp}, \omega^{\perp} \rangle .$$

If follows that one of ω' and $\omega^{\perp}$ is zero, i.e. ω lies either in R' or in $R^{\perp}$. We see that R' and $R^{\perp}$ are T-subgroups in R.

Q.E.D.

3.9. Fix $n \geq 0$. We apply Lemma 3.8 to $\{e_1, \ldots, e_p\} = \{x_\lambda \mid \lambda \in \mathcal{P}_n\}$. More precisely, let $\lambda_1, \ldots, \lambda_p$ be all elements of $\mathcal{P}_n$ ordered in such a way that for $i < j$ $c.f(\lambda_j^t)$ is lexicographically higher than $c.f(\lambda_i^t)$. Put $e_i = x_{\lambda_i}$. We must prove that $e_1, \ldots, e_p$ form a $\mathbb{Z}$-basis in R_n. Let us derive this from the equality

(*) $\det (\langle e_i, e_j \rangle) = 1$

Indeed, according to Lemma 3.8, (*) implies that $\{e_1, \ldots, e_p\}$ is a basis of a certain T-subgroup $R' \subset R_n$. To prove that $R' = R_n$ it suffices to verify that no irreducible element $\omega \in R_n$ can be orthogonal to all e_i. But this is clear since

$$\langle \omega, e_p \rangle = \langle \omega, x_{(1^n)} \rangle = \langle \omega, \rho^n \rangle \neq 0$$

It is easy to calculate the inner products $\langle e_i, e_j \rangle$ explicitly (see 3.17 (c) below), but a direct computation of $\det (\langle e_i, e_j \rangle)$ seems to be rather tedious. So, in order to prove (*) we apply the following trick due to J.N.Bernstein. Put

$$e_i' = y_{\lambda_i^t} \qquad (i = 1, \ldots, p) .$$

We shall derive (*) from the following two statements.

(a) The matrix $(\langle e_i, e_j' \rangle)$ is triangular with 1's down the diagonal; in particular, its determinant equals 1.

(b) All e_j' are integral linear combinations of the elements (e_i).

Indeed, by (b)

$$e'_j = \sum_{i=1}^{p} a_{ij}\, e_i \quad , \text{ where } \quad a_{ij} \in \mathbb{Z}$$

Clearly, the matrix $(\langle e_i, e'_j \rangle)$ is a product of the Gram matrix $(\langle e_i, e_j \rangle)$ and the matrix (a_{ij}). According to (a), one has:

$$1 = \det(\langle e_i, e'_j \rangle) = \det(\langle e_i, e_j \rangle) \cdot \det(a_{ij}) .$$

Since the factors on the right-hand side are both integers, each of them equals ± 1. Since the Gram determinant is always non-negative, we see that $\det(\langle e_i, e_j \rangle) = 1$, as desired.

Summarizing, we see that the equality

$$R = \mathbb{Z}[x_1, x_2, \ldots]$$

is derived from the assertions (a) and (b). The assertion (a) is proved in 3.10 while (b) in 3.11-3.12.

3.10. The assertion 3.9 (a) follows at once from the next.

Proposition. If $\lambda, \mu \in \mathcal{P}$ and c.f. (μ) is lexicographically higher than c.f. (λ^t) then

$$y^*_\mu (x_\lambda) = 0 .$$

Furthermore, $y^*_{\lambda^t}(x_\lambda) = 1$ for all $\lambda \in \mathcal{P}$.

Proof. Let $r(\lambda) = r$. The formulas 3.6 $(*)$ and 3.1 (c) imply that

$$y^*_r (x_\lambda) = x_{\lambda^\leftarrow} \qquad \text{and}$$

$$y^*_m (x_\lambda) = 0 \qquad \text{for} \quad m > r \qquad \text{(see 3.7).}$$

Now apply 3.7 (1).

Q. E.D.

3.11. The assertion 3.9 (b) follows easily from the relations 3.1 (e) between (x_n) and (y_n). Indeed, these formulas can be rewritten as

$$y_n = \sum_{k \geqslant 1} (-1)^{k-1} x_k y_{n-k} \qquad (n \geqslant 1)$$

Using induction on n one obtaims that all y_n (and hence all y_λ) are polynomials in $x_1, x_2, \ldots$ with integral coefficients, as desired.

To prove the relations 3.1 (e) we shall use the important notion of the conjugation of a Hopf algebra (see [8], § 8, or Appendix 1, Proposition A1.6 below). In the theory of PSH-algebras it is convenient to replace the conjugation $T : R \to R$ by the morphism $t : R \to R$ differing from T on R_n by the multiple $(-1)^n$.

Proposition. (a) The map t is a (positive) involutive automorphism of the PSH-algebra R.

(b) For all $n \in \mathbb{Z}$ we have

$$t(x_n) = y_n, \quad t(y_n) = x_n \quad .$$

Proof. (a) Clearly, t as well as T is an involutive automorphism of the Hopf algebra R (see Appendix 1, Prop. A1.6). It remains to verify that t takes irreducible elements to irreducible ones. Let us prove that T (and hence t) is an isometry of R. We recall that T is uniquely determined by the property of commutativity of the diagram:

$$\text{(D)} \qquad \begin{array}{ccccccc} R & \xrightarrow{m^*} & R\otimes R & \xrightarrow{id\otimes T} & R\otimes R & \xrightarrow{m} & R \\ & \searrow^{e^*} & & \mathbb{Z} & & \nearrow_{e} & \end{array}$$

Consider the diagram (D^*) obtained by passing to adjoint operators in (D). The axiom (S) implies that (D^*) is obtained from (D) by replacing T by the adjoint operator T^* (T^* exists since each R_n has finite rank). It follows that

$$T = T^* .$$

Since $T^2 = id$, we see that

$$T^* = T = T^{-1}$$

hence T is an isometry, as desired.

Clearly, an element $\omega \in R_n$ is irreducible if and only if $\langle\omega,\omega\rangle = 1$ and $\langle\omega, \rho^n\rangle > 0$. By definition of T we have $T(\rho) = -\rho$ hence $t(\rho) = \rho$, so $t(\rho^n) = \rho^n$ for $n \geqslant 0$. Therefore, if $\omega \in R_n$ is irreducible then

$$\langle t(\omega), t(\omega)\rangle = \langle\omega,\omega\rangle = 1,$$

and

$$\langle t(\omega), \rho^n\rangle = \langle t(\omega), t(\rho^n)\rangle = \langle\omega, \rho^n\rangle > 0$$

hence $t(\omega)$ is irreducible, as desired.

(b) By definition of T, we have

$$m \circ (id \otimes T) \circ m^* (x_2) = 0.$$

Apply 3.1 (d):

$$0 = m \circ (id \otimes T)\ (1 \otimes x_2 + \rho \otimes \rho + x_2 \otimes 1) = T(x_2) - \rho^2 + x_2.$$

Therefore,

$$t(x_2) = T(x_2) = \rho^2 - x_2 = y_2.$$

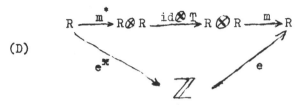

From the fact that t is an isometry, we easily derive that

$$(*) \qquad (t(a))^* = t \circ a^* \circ t^{-1} \qquad (a \in R).$$

Hence

$$(x_2^* \circ t)\,(x_n) = (t \circ y_2^*)\,(x_n) = 0 \quad \text{for} \quad n \geq 0.$$

By 3.1 (b) this implies that $t(x_n) = y_n$ for all $n \geq 0$. Similarly, $t(y_n) = x_n$.

Q.E.D.

3.12. The end of the proof of 3.1 (e). By definition of T,

$$m \circ (\mathrm{id} \otimes T) \circ m^* \,(x_n) = 0 \quad \text{for} \quad n \geq 1.$$

Applying 3.1 (d) and 3.11 (b), one obtains the formulas 3.1 (e). We have already derived the assertion

$$R = \mathbb{Z}[x_1, x_2, \ldots]$$

from these formulas (see 3.11). The proof of the equality

$$R = \mathbb{Z}[y_1, y_2, \ldots]$$

is quite similar; another way is to apply the automorphism t.

Q.E.D.

3.13. Proof of 3.1 (f). The existence of t is proved in 3.11; it remains to prove the uniqueness. Let $\tau : R \longrightarrow R$ be an automorphism of the PSH-algebra R. According to 3.1 (e), it suffices to prove that either $\tau(x_n) = x_n$ for all n, or $\tau(x_n) = y_n$ for all n. But this assertion follows at once from Proposition 3.6 applied to the form $\delta = \delta_x \circ \tau$.

Q.E.D.

3.14. Proof of 3.1 (g). Write R' as a polynomial algebra $\mathbb{Z}[x_1', x_2', \ldots]$ according to 3.1 (e). Consider the ring homomorphism $j : R \to R'$ which sends x_n to x_n'. By 3.1 (d), j is a Hopf algebra morphism. We must prove that j takes irreducible elements to irreducible ones. It is easy to see that the formulas 3.1 (c) and 3.6 (*) allow one to compute the inner products

$$\langle x_\lambda , x_\mu \rangle$$

explicitly(we will do this later in 3.17). This implies that j is an isometry. Now apply the arguments similar to those at the end of the proof of 3.11 (a).

Q.E.D.

Theorem 3.1 is completely done.

3.15. Let us find all primitive elements of R.

Proposition. (a) For any $n \geqslant 1$ there exists the unique primitive element $z_n \in R_n$ such that

$$\langle z_n, x_n \rangle = 1 .$$

The subgroup of primitive elements in R_n equals $\mathbb{Z} \cdot z_n$.

(b) For each field K of characteristic 0 the Hopf algebra $R_K = R \otimes K$ equals

$$K[z_1, z_2, \ldots] .$$

Proof. (a). According to 1.7 and 3.1(e), an element $z \in R_n$ is primitive if and only if it is orthogonal to all x_λ except x_n. Clearly, any such element is proportional to the one with the least positive value of $\langle z, x_n \rangle$. Since the Gram determinant

$$\det (\langle x_\lambda , x_\mu \rangle)_{\lambda, \mu \in \mathcal{P}_n}$$

equals 1 (see 3.9), this latter value is 1. Our assertion follows.

(b). This follows immediately from (a) and the Theorem quoted in 2.1 (see also Appendix 1, A1.1 and A1.2).

Q.E.D.

3.16. For any $\lambda = (l_1,\dots,l_r) \in \mathcal{P}$ where $l_i \in \mathbb{N}$ put

$$z_\lambda = z_{l_1} z_{l_2} \dots z_{l_r} ;$$

put also $z_\emptyset = 1$. By 3.15 (b), the elements z_λ form a $\mathbb{Q}$-basis in $R_{\mathbb{Q}} = R \otimes \mathbb{Q}$. Let us describe the relations between this basis and each of the bases (x_λ) and (y_λ). We shall use the method of generating functions.

Proposition. Let $R[[\xi]]$ be the ring of formal power series in one indeterminate ξ over R. Define the elements $X(\xi)$, $Y(\xi)$, and $Z(\xi)$ of $R[[\xi]]$ by

$$X(\xi) = \sum_{n\geq 0} x_n \cdot \xi^n, Y(\xi) = \sum_{n\geq 0} y_n \cdot \xi^n , \quad Z(\xi) = \sum_{n\geq 0} z_{n+1} \cdot \xi^n .$$

Then in $R[[\xi]]$ one has

$$Y(\xi) = (X(-\xi))^{-1} \text{ and}$$

$$Z(\xi) = \frac{d}{d\xi} \ln X(\xi) = (\frac{d}{d\xi} X(\xi)) / X(\xi) .$$

(clearly these formulas allow one to write down the elements y_λ and z_λ in the basis (x_λ)).

Proof. The first equality is equivalent to the formulas 3.1 (e). Let us prove the second one. Denote temporarily by

$$\tilde{Z}(\xi) = \sum_{n\geq 0} \tilde{z}_{n+1} \cdot \xi^n$$

the power series

$$\frac{d}{d\xi} \ln X (\xi) = \frac{d}{d\xi} X(\xi) \ / \ X(\xi) .$$

Since the constant term of $X(\xi)$ equals $x_0 = 1$, the series $X(\xi)$

is invertible in $R[[\xi]]$ hence $\tilde{Z}(\xi) \in R[[\xi]]$. Clearly, $\tilde{z}_n \in R_n$ for $n \geqslant 1$. It remains to prove that all elements $\tilde{z}_n$ are primitive and satisfy $\langle \tilde{z}_n, x_n \rangle = 1$ (see 3.15 (a)).

We shall apply the following assertion which can be verified directly.

(*) For any ring homomorphism $\varphi : \mathcal{A} \longrightarrow \mathcal{B}$ the corresponding morphism $\mathcal{A}[[\xi]] \longrightarrow \mathcal{B}[[\xi]]$ obtained by applying φ coefficientwise (it will be denoted also by φ) commutes with the operator of logarithmic derivative.

First we apply (*) to the homomorphism

$$m^* : R \longrightarrow R \otimes R.$$

For any $U = \sum u_n \xi^n \in R[[\xi]]$ denote by U_1 (U_2) the series

$$U_1 = \sum (u_n \otimes 1) \xi^n \qquad (U_2 = \sum (1 \otimes u_n) \xi^n)$$

in $(R \otimes R)[[\xi]]$. Under this notation one can rewrite the assertion 3.1 (d) as

$$m^* X(\xi) = X_1(\xi) \cdot X_2(\xi) .$$

Therefore,

$$m^* \tilde{Z}(\xi) = m^* \left[\frac{d}{d\xi} \ln X(\xi) \right] = \frac{d}{d\xi} \ln m^* X(\xi) =$$

$$= \frac{d}{d\xi} \ln \left[X_1(\xi) \cdot X_2(\xi) \right] = \frac{d}{d\xi} \ln X_1(\xi) + \frac{d}{d\xi} \ln X_2(\xi) =$$

$$= \tilde{Z}_1(\xi) + \tilde{Z}_2(\xi) .$$

This means that all coefficients $\tilde{z}_n$ are primitive.

Now apply the assertion (*) to the homomorphism

$$\delta_x : R \longrightarrow \mathbb{Z} \qquad \text{(see 3.6).}$$

By definition,

$$\delta_x X(\xi) = \sum_{n\geqslant 0} \xi^n = (1-\xi)^{-1} .$$

Therefore,

$$\delta_x \widetilde{Z}(\xi) = \delta_x\left[\frac{d}{d\xi} \ln X(\xi)\right] = \frac{d}{d\xi} \ln\left[\delta_x X(\xi)\right] =$$

$$= \frac{d}{d\xi} \ln (1-\xi)^{-1} = (1-\xi)^{-1} = \sum_{n\geqslant 0} \xi^n .$$

This means that $\delta_x(\widetilde{z}_n) = 1$ for $n \geqslant 1$, i.e. $\langle x_n, \widetilde{z}_n \rangle = 1$, as desired. Q.E.D.

3.17. We conclude this section by computing the inner products between all elements (x_λ), (y_λ), and (z_λ). To interpret these results consider the dual bases $(x_\lambda^\perp)$, $(y_\lambda^\perp)$ and $(z_\lambda^\perp)$ defined by

$$\langle x_\lambda^\perp , x_\mu \rangle = \langle y_\lambda^\perp , y_\mu \rangle = \langle z_\lambda^\perp , z_\mu \rangle = \delta_{\lambda,\mu}$$

According to 3.9, each of the sets $\{x_\lambda^\perp \mid \lambda \in \mathcal{P}\}$ and $\{y_\lambda^\perp \mid \lambda \in \mathcal{P}\}$ is a $\mathbb{Z}$-basis of R; the set $\{z_\lambda^\perp \mid \lambda \in \mathcal{P}\}$ is a $\mathbb{Q}$-basis of $R_{\mathbb{Q}}$. Clearly, the inner products of a vector with all elements of certain basis are its coordinates with respect to the dual basis. Thus, we shall compute the transition matrices between each of the bases (x_λ), (y_λ) and (z_λ) and each of the dual bases.

Proposition. (a) For any $\lambda \in \mathcal{P}$ one has

$$t(x_\lambda) = y_\lambda , \; t(y_\lambda) = x_\lambda , \; t(z_\lambda) = (-1)^{|\lambda| - r(\lambda)} \bar{z}_\lambda$$

(see 3.7, 3.11).

(b) For $n \geq 1$ one has

$$X^* z_n = z_n + 1, \quad Y^* z_n = z_n + (-1)^{n-1} \quad \text{(see 3.6)}$$

The operator $z_n^* : R \to R$ is a derivation of R and acts on the generators x_k and y_k by

$$z_n^* (x_k) = x_{k-n} \quad , \quad z_n^* (y_k) = (-1)^{n-1} y_{k-n} \quad .$$

(c) Let $\lambda = (l_1,\ldots,l_r)$ and $\mu = (m_1,\ldots,m_s)$ be two partitions with all l_i and m_j being non-zero. Denote by $M^{\lambda,\mu}$ the set of $r \times s$ - matrices over $\mathbb{Z}^+$ such that the sums of their entries over the rows equal $l_1,\ldots,l_r$ while the sums over columns are $m_1,\ldots,m_s$. Then

$$< x_\lambda , x_\mu > = < y_\lambda , y_\mu > = |M^{\lambda,\mu}| ,$$

while $< x_\lambda, y_\mu >$ is equal to the number of matrices from $M^{\lambda,\mu}$ with entries 0 and 1. Furthermore, $< x_\mu , z_\lambda >$ equals the number of matrices from $M^{\lambda,\mu}$, which have only one non-zero entry in each row, while

$$< y_\mu , z_\lambda > = (-1)^{|\lambda|-r(\lambda)} \cdot < x_\mu , z_\lambda > .$$

Finally, $< z_\lambda , z_\mu > = 0$ unless $\lambda = \mu$, while

$$< z_\lambda , z_\lambda > = \prod_{k \geq 1} k^{r_k(\lambda)} \cdot r_k(\lambda) ! \qquad \text{(see 3.7)}$$

Proof. (a) The first two formulas are obvious. To prove the last one it suffices to verify that

$$t(z_n) = (-1)^{n-1} z_n \ .$$

i.e. that $T(z_n) = -z_n$. This follows at once from the primitivity of z_n and the definition of a conjugation T.

(b) By 1.9 (g), for $0 < k < n$

$$x_k^*(z_n) = y_k^*(z_n) = 0 \;;$$

by 1.9 (a) and definition of z_n,

$$x_n^*(z_n) = \langle x_n, z_n \rangle = 1 \,.$$

Furthermore,

$$y_n^*(z_n) = \langle y_n, z_n \rangle = \langle t(y_n), t(z_n) \rangle = (-1)^{n-1} \langle x_n, z_n \rangle = (-1)^{n-1}.$$

We have proved the first two formulas. The assertion that z_n^* is a derivation, follows from 1.9 (f).

One can easily derive from 3.1 (b) that for $a \in R_n$ the element $a^*(x_k)$ is proportional to x_{k-n}. In particular,

$$z_n^*(x_k) = c \cdot x_{k-n} \,.$$

We have

$$c = x_{k-n}^* \circ z_n^*(x_k) = z_n^* \circ x_{k-n}^*(x_k) = z_n^*(x_n) = 1,$$

so $z_n^*(x_k) = x_{k-n}$. The proof of the formula for $z_n^*(y_k)$ is quite similar.

(c) Without loss of generality we assume that $|\lambda| = |\mu|$. By 1.9 (a), (c):

$$\langle x_\lambda, x_\mu \rangle = x_\lambda^*(x_\mu) = x_{l_1}^* \circ \ldots \circ x_{l_r}^*(x_\mu)$$

(the similar holds for $\langle y_\lambda, x_\mu \rangle = y_\lambda^*(x_\mu)$ and $\langle z_\lambda, x_\mu \rangle = z_\lambda^*(x_\mu)$). Apply (b), 3.1 (c) and 3.6 (*):

$$x_l^*(x_\mu) = \sum x_{(m_1-k_1, m_2-k_2, \ldots, m_s-k_s)} \quad (k_i \in \mathbb{Z}^+;\ k_1+\ldots+k_s=l)$$

$$y_l^*(x_\mu) = \sum x_{(m_1-k_1, m_2-k_2, \ldots, m_s-k_s)} \quad (k_i=0,1; k_1+\ldots+k_s=l)$$

$$z_1^*(x_\mu) = \sum_{i=1}^{s} x_{(m_1,m_2,\ldots,m_i-1,\ldots,m_s)} \quad .$$

Iterating, we obtain the desired values for $\langle x_\lambda, x_\mu \rangle$, $\langle y_\lambda, x_\mu \rangle$ and $\langle z_\lambda, x_\mu \rangle$. Now use (a) and 3.11 (a):

$$\langle y_\lambda, y_\mu \rangle = \langle t(x_\lambda), t(x_\mu) \rangle = \langle x_\lambda, x_\mu \rangle ;$$

$$\langle y_\mu, z_\lambda \rangle = \langle t(x_\mu), (-1)^{|\lambda|-r(\lambda)} t(z_\lambda) \rangle = (-1)^{|\lambda|-r(\lambda)} \langle x_\mu, z_\lambda \rangle .$$

The fact that all z_λ are mutually orthogonal, follows from Proposition 2.3 with the account of Remark 2.4. Furthermore, the computations in 2.3 show that

$$\langle z_\lambda, z_\lambda \rangle = \prod_{k\geq 1} r_k ! \cdot \prod_{i=1}^{r} \langle z_{l_i}, z_{l_i} \rangle$$

It remains to verify that

$$\langle z_n, z_n \rangle = n \ .$$

It follows from 3.15 that $\langle z_n, z_n \rangle$ equals the coefficient of x_n in the decomposition of z_n in the basis (x_λ). By 3.16:

$$\sum_{n\geq 1} z_n \xi^{n-1} = \left[\frac{d}{d\xi} X(\xi)\right] / X(\xi) = \left(\sum_{n\geq 1} n x_n \xi^{n-1}\right)(1+x_1\xi + x_2\xi^2 + \ldots)^{-1}$$

This implies that our coefficient equals n, as desired.

Q.E.D.

3.18. Corollary. For any $\lambda = (1^{r_1}, 2^{r_2}, \ldots) \in \mathcal{P}$ put

$$c_\lambda = \prod_{k\geq 1} k^{r_k} \cdot r_k !$$

Then $z_\lambda^\perp = \frac{1}{c_\lambda} z_\lambda$.

§ 4. Universal PSH-algebra: irreducible elements.

4.1. We retain the notation of § 3. Thus, R is a PSH-algebra with the unique irreducible primitive element ρ . We have two $\mathbb{Z}$-bases (x_λ) and (y_λ) in R and the $\mathbb{Q}$- basis (z_λ) in $R \otimes \mathbb{Q}$, parametrized by partitions $\lambda \in \mathcal{P}$. In this section we shall study the basis Ω in R, consisting of irreducible elements. We shall compute explicitly the transition matrices between Ω and all bases from § 3. First we show that irreducible elements of R are also naturally parametrized by partitions.

Proposition-definition. For any $\lambda \in \mathcal{P}$ there exists the unique irreducible element $\{\lambda\}$ in $R_{|\lambda|}$ such that $x_\lambda^*(\{\lambda\}) \neq 0$ and $y_{\lambda^t}^*(\{\lambda\}) \neq 0$. We have

$$x_\lambda^*(\{\lambda\}) = y_{\lambda^t}^*(\{\lambda\}) = 1.$$

In other words, $\{\lambda\}$ can be characterized as a unique common irreducible constituent of x_λ and y_{λ^t}, and it satisfies

$$\langle \{\lambda\}, x_\lambda \rangle = \langle \{\lambda\}, y_{\lambda^t} \rangle = 1 .$$

The map $\lambda \longmapsto \{\lambda\}$ is a bijection between $\mathcal{P}$ and Ω.

Proof. The condition $x_\lambda^*(\{\lambda\}) \neq 0$, clearly, means that $\{\lambda\}$ is an irreducible constituent of x_λ. We know that $y_{\lambda^t}^*$ is positive and

$$y_{\lambda^t}^* \quad (x_\lambda) = 1 \quad \text{(see 1.9.(d) and 3.10)}.$$

This implies the existence and uniqueness of $\{\lambda\}$ and the equalities

$$x_\lambda^*(\{\lambda\}) = \langle x_\lambda, \{\lambda\} \rangle = y_{\lambda^t}^*(\{\lambda\}) = 1 .$$

Now let $\lambda, \mu \in \mathcal{P}$ and $\lambda \neq \mu$; we can assume that c.f.(μ^t)

is lexicographically higher than c.f. (λ^t) (see 3.7). Applying again the positivity of $y^*_{\mu^t}$ and Proposition 3.10, we obtain that $y^*_{\mu^t}(\{\lambda\}) = 0$ hence $\{\lambda\} \neq \{\mu\}$.

Finally, by 3.1 (e)

$$rk_{\mathbf{Z}} R_n = |\mathcal{P}_n| ;$$

this implies that the elements $\{\lambda\}$ exhaust all irreducible elements of R.

Q.E.D.

4.2. Corollary. For all $\lambda \in \mathcal{P}$

$$t(\{\lambda\}) = \{\lambda^t\} .$$

Proof. Apply 3.11 and 3.17 (a)

Q.E.D.

4.3. Now we compute the action of the operators X^* and Y^* (see 3.6) on irreducible elements. This result will play the crucial role in the sequel (see 4.4). Let us give two combinatorial definitions.

Let $\lambda, \mu \in \mathcal{P}$ be two Young diagrams (see 3.7). We write

$$\mu \dashv \lambda$$

if $\lambda^{\leftarrow} \subseteq \mu \subseteq \lambda$. This means that μ is obtained by removing at most one point from each row of λ. We write

$$\mu \perp \lambda$$

instead of $\mu^t \dashv \lambda^t$; clearly, the relation „$\mu \perp \lambda$" is equivalent to „$\lambda^{\uparrow} \subseteq \mu \subseteq \lambda$" and means geometrically that μ is obtained by removing at most one point from each column of λ.

Theorem. For all $\lambda \in \mathcal{P}$ we have

$$X^*(\{\lambda\}) = \sum_{\mu \perp \lambda} \{\mu\} \quad \text{and} \quad Y^*(\{\lambda\}) = \sum_{\mu \dashv \lambda} \{\mu\}$$

This is proved in 4.5–4.10.

4.4. The next proposition shows the significance of Theorem 4.3.

Proposition. For each $n \geq 1$ the operators

$$X^* - 1 = \sum_{k \geq 1} x_k^* \quad \text{and} \quad Y^* - 1 = \sum_{k \geq 1} y_k^*$$

are injective on R_n.

Proof. It suffices to consider only $(X^* - 1)$. Passing to adjoint operators, we see that our assertion is equivalent to the following:

for $n \geq 1$ the operator from $\bigoplus_{1 \leq k \leq n} R_{n-k}$ to R_n,

which acts on R_{n-k} by multiplying by x_k, is an epimorphism.

But this follows at once from 3.1 (e)

Q.E.D.

This proposition allows us to reduce the verification of various identities in R_n to that in R_k for $k < n$, by applying the operator $X^* - 1$ (or $Y^* - 1$). Therefore, Theorem 4.3 allows us to prove various relations between irreducible elements of degree n by induction on n. For example, we obtain the following useful assertion:

(*) If we assign to each $\lambda \in \mathcal{P}$ an element $[\lambda] \in R_{|\lambda|}$ in such a way that $[\emptyset] = 1$ and

$$X^*([\lambda]) = \sum_{\mu \perp \lambda} [\mu] \quad \text{for all } \lambda \in \mathcal{P}$$

then

$$[\lambda] = \{\lambda\} \quad \text{for all } \lambda \in \mathcal{P}.$$

4.5. We begin to prove Theorem 4.3. Let

$$\mathrm{c.f.}(\lambda) = (l_1,\dots,l_r),$$

where $r = r(\lambda)$ (see 3.7). First, we want to prove that

$$(*) \qquad y_r^*(\{\lambda\}) = \{\lambda^{\leftarrow}\},$$

a main special case of Theorem 4.3.

Set $u = y_r^*(\{\lambda\})$. By 1.9 (d) and 3.10:

$$0 \leq u \leq y_r^*(x_\lambda) = x_{\lambda^{\leftarrow}} \ ;$$

hence for any irreducible constituent ω of u

$$x_{\lambda^{\leftarrow}}^*(\omega) \neq 0.$$

On the other hand, we have

$$y_{(\lambda^{\leftarrow})^t}^*(u) = y_{(\lambda^{\leftarrow})^t}^* \circ y_r^*(\{\lambda\}) = y_{\lambda^t}^*(\{\lambda\}) = 1.$$

It follows that $\{\lambda^{\leftarrow}\}$ is an irreducible constituent of u. Thus, to prove $(*)$ we must only verify that u is irreducible, i.e. $\langle u,u\rangle = 1$. Since $0 \leq u \leq x_{\lambda^{\leftarrow}}$, it suffices to verify

$$(**) \qquad x_{\lambda^{\leftarrow}}^* \circ y_r^*(\{\lambda\}) = 1 .$$

This equality is proved in 4.6–4.8.

4.6. Lemma. Let $p,q \in \mathbb{N}$ and $v \in R$. Suppose that $x_l^*(v) = 0$ for $l > p$, and $y_m^*(v) = 0$ for $m > q$. Then

$$x_p^* \circ y_q^*(v) = 0 \quad \text{and} \quad x_{p-1}^* \circ y_q^*(v) = x_p^* \circ y_{q-1}^*(v).$$

Proof. Let us pass to adjoint operators in the formulas 3.1 (e). We obtain that

$$\sum_{k=0}^{n} (-1)^k x_k^* \circ y_{n-k}^* = 0 \qquad \text{for } n \geq 1.$$

It remains to apply this operator equality to v for $n = p+q$ and $n = p+q-1$.

Q.E.D.

4.7. Reverting to the notation of 4.5, we put

$$v_k = x^*_{l_k} \circ x^*_{l_{k-1}} \circ \dots \circ x^*_{l_1}(\{\lambda\}) \qquad (k = 0,1,\dots,r)$$

Lemma. For any $k = 0,1,\dots,r$ we have

$$x^*_l(v_k) = 0 \quad \text{for } l > l_{k+1} \quad \text{and}$$

$$y^*_m(v_k) = 0 \quad \text{for } m > r-k.$$

Proof. The first statement follows from the fact that $x^*_\mu(y_{\lambda^t}) = 0$ when c.f.(μ) is lexicographically higher than c.f.(λ^t) (one has to interchange x and y in 3.10). To prove the second assertion we proceed by induction on k.

When $k = 0$ we must prove that $y^*_m(\{\lambda\}) = 0$ for $m > r$. This follows at once from the equality $y^*_m(x_\lambda) = 0$ (see 3.10). Let now $k > 0$, and assume that we have already proved that

$$y^*_m(v_{k-1}) = 0 \quad \text{for } m > r-k+1.$$

Clearly,

$$y^*_m(v_k) = x^*_{l_k} \circ y^*_m(v_{k-1}),$$

so $y^*_m(v_k) = 0$ for $m > r-k+1$. It remains to prove that

$$x^*_{l_k} \circ y^*_{r-k+1}(v_{k-1}) = 0.$$

It suffices to apply Lemma 4.6 to $v = v_{k-1}$.

Q.E.D.

4.8. Now we apply Lemma 4.6 to $v = v_k$ taking into account Lemma 4.7. We obtain for $0 \leq k < r$:

$$x_{1_{k+1}-1} \circ y^*_{r-k}(v_k) = y^*_{r-k-1} \circ x^*_{1_{k+1}}(v_k) = y^*_{r-k-1}(v_{k+1}).$$

Using these equalities successuvely for $k = 0,1,\ldots,r-1$, we obtain:

$$x^*_{\lambda^{\leftarrow}} \circ y^*_r(\{\lambda\}) = v_r = x^*_{\lambda}(\{\lambda\}) = 1.$$

The formula (**) from 4.5, and hence the equality

$$y^*_r\{\lambda\} = \{\lambda^{\leftarrow}\}$$

are proved.

4.9. Now we consider the general case of Theorem 4.3.It suffi ces to prove only the first equality.

Define the numbers $a_i(\lambda, \mu)$ where $\lambda, \mu \in \mathcal{P}$, by

$$x^*_i(\{\lambda\}) = \sum_{\mu} a_i(\lambda,\mu) \cdot \{\mu\} .$$

We must prove that

$$a_i(\lambda,\mu) = \begin{cases} 1 & \text{if } \mu \perp \lambda \text{ and } |\lambda| - |\mu| = i \\ 0 & \text{otherwise} \end{cases}$$

4.10. Lemma. Let $\lambda, \mu \in \mathcal{P}$. Then

(a) $\mu \perp \lambda$ if and only if $\mu^{\leftarrow} \perp \lambda^{\leftarrow}$ and $r(\mu) \leq r(\lambda) \leq r(\mu) + 1$.

(b) If $a_i(\lambda,\mu) > 0$ then $r(\mu) \leq r(\lambda) \leq r(\mu) + 1$.

(c) If $r(\lambda) = r(\mu)$ then $a_i(\lambda,\mu) = a_i(\lambda^{\leftarrow}, \mu^{\leftarrow})$.

(d) If $r(\lambda) = r(\mu) + 1$ then $a_i(\lambda,\mu) = a_{i-1}(\lambda^{\leftarrow}, \mu^{\leftarrow})$.

Proof. Part (a) follows at once from definitions. One has only to observe that $r(\lambda)$ is the length of the first column of the Young diagram λ , and $\lambda^{\leftarrow}$ is obtained by removing the first column from λ.

(b) Clearly,

$$a_i(\lambda,\mu)>0 \Longleftrightarrow \{\mu\} \leq x_i^*(\{\lambda\}) \Longleftrightarrow \{\lambda\} \leq x_i \cdot \{\mu\} .$$

Therefore, if $r > r(\lambda)$ then

$$0 \leq y_r^* (\{\mu\}) \leq y_r^* \circ x_i^* (\{\lambda\}) = x_i^* \circ y_r^* (\{\lambda\}) = 0,$$

hence $r(\mu) \leq r(\lambda)$.

On the other hand, if $r > r(\mu) + 1$ then

$$0 \leq y_r^*(\{\lambda\}) \leq y_r^*(x_i \cdot \{\mu\}) = x_i \cdot y_r^*(\{\mu\}) + x_{i-1} \cdot y_{r-1}^*(\{\mu\}) = 0$$

(see 3.1 (c) and 3.6 (*))
hence $r(\lambda) \leq r(\mu) + 1$

(c) Let $r = r(\lambda)$. By 4.5 (*), $y_r^* (\{\lambda\}) = \{\lambda^{\leftarrow}\}$, so

$$x_i^* (\{\lambda^{\leftarrow}\}) = x_i^* \circ y_r^* (\{\lambda\}) = \sum_{\mu} a_i(\lambda,\mu)\, y_r^* (\{\mu\}).$$

According to (b), if $a_i(\lambda,\mu) > 0$ then $r(\mu) \leq r$, so we have

$$x_i^* (\{\lambda^{\leftarrow}\}) = \sum_{r(\mu)=r} a_i(\lambda,\mu) \cdot \{\mu^{\leftarrow}\}$$

It is easy to see that a partition μ is uniquely determined by the number $|\mu|$ and the partition $\mu^{\leftarrow}$. It follows that $a_i(\lambda,\mu) = a_i(\lambda^{\leftarrow},\mu^{\leftarrow})$, as desired.

(d) Let now $r = r(\mu) + 1$. Using 3.1 (c), 3.6 and 4.5 (*) we obtain that

$$y_r^* (x_i \cdot \{\mu\}) = x_{i-1} \cdot \{\mu^{\leftarrow}\} .$$

Hence

$$x_{i-1}\cdot\{\mu^{\leftarrow}\} = \sum_{\lambda} a_i(\lambda,\mu)\cdot y_r^*(\{\lambda\}).$$

According to (b), if $a_i(\lambda,\mu)>0$ then $r(\lambda)\leq r$, so we have

$$x_{i-1}\cdot\{\mu^{\leftarrow}\} = \sum_{r(\lambda)=r} a_i(\lambda,\mu)\cdot\{\lambda^{\leftarrow}\}.$$

The same argument as in (c) shows that $a_i(\lambda,\mu) = a_{i-1}(\lambda^{\leftarrow},\mu^{\leftarrow})$.

Q.E.D.

Theorem 4.3 follows at once from this Lemma by means of induction on $|\lambda|$.

4.11. Before investigating the relations between the basis Ω and the bases of § 3, we will extend Ω to a larger class of positive elements in R, parametrized by so called skew diagrams.

For any $\lambda,\mu\in\mathcal{P}$ put

$$\{\lambda\setminus\mu\} = \{\mu\}^*(\{\lambda\});$$

this notation is justified by the fact that the element $\{\lambda\setminus\mu\}$ is non-zero only when $\lambda\supset\mu$ (as Young diagrams) and depends only on the set-theoretic difference between the diagrams λ and μ (this will be proved in 4.13).

By 1.9 (d), all elements $\{\lambda\setminus\mu\}$ are positive. By 1.9 (a):

$$\{\lambda\setminus\mu\}\in R_{|\lambda|-|\mu|}, \qquad \{\lambda\setminus\emptyset\}=\{\lambda\}$$

and for $|\lambda|=|\mu|$ we have

$$\{\lambda\setminus\mu\}=\delta_{\lambda\mu}.$$

Finally, we derive from 1.9 (b) that

$$(*)\qquad m^*(\{\lambda\}) = \sum_{\mu\in\mathcal{P}}\{\lambda\setminus\mu\}\otimes\{\mu\}.$$

4.12. Let us introduce some combinatorial terminology. Define the partial order „ $\leq_p$ " on $\mathbb{N}\times\mathbb{N}$ by

$$(i,j)\leq_p(i',j') \iff i\leq i',\ j\leq j'.$$

<u>A skew diagram</u> is a finite subset $\mathscr{X}\subset\mathbb{N}\times\mathbb{N}$ which is convex with respect to this order, i.e. such that

$$a,b\in\mathscr{X}\ ,\ a\leq_p c\leq_p b \implies c\in\mathscr{X}.$$

Clearly, a (non-empty) Young diagram can be defined as a skew diagram, containing the point (1,1). It is easy to see that the difference $\lambda\setminus\mu$, where $\lambda\supset\mu$ are two Young diagrams, is a skew diagram; conversely, any skew diagram has such a form. As in 3.7, we write $|\mathscr{X}|$ for the number of points of $\mathscr{X}$, and $\mathscr{X}^t$ for the transposed diagram.

Let $\mathscr{X}$ be a skew diagram. A subset $\mathscr{X}'\subset\mathscr{X}$ is <u>regular</u> if it contains with each point a all points of $\mathscr{X}$, which are greater than a (i.e.

$$a\in\mathscr{X}',\ b\in\mathscr{X},\ a<_p b \implies b\in\mathscr{X}').$$

Clearly, if $\mathscr{X}'\subset\mathscr{X}$ is regular then $\mathscr{X}'$ and $\mathscr{X}\setminus\mathscr{X}'$ are skew diagrams. We shall write $\mathscr{X}'\dashv\mathscr{X}$ ($\mathscr{X}'\perp\mathscr{X}$) if $\mathscr{X}'$ is obtained from $\mathscr{X}$ by removing a regular subset which has at most one point in each row (column).

Obviously these definitions are compatible with those given in 4.3 for Young diagrams. Moreover, if $\mathscr{X}=\lambda\setminus\mu$ is represented as a difference between two Young diagrams $\lambda\supset\mu$ then skew diagrams $\mathscr{X}'\dashv\mathscr{X}$ are just those of the form $\lambda'\setminus\mu$ where

$$\mu\subset\lambda'\dashv\lambda$$

(the similar holds for the relation „ $\perp$ ").

4.13. Proposition. Let λ and μ be two Young diagrams. The element $\{\lambda\setminus\mu\}$ is non-zero if and only if $\lambda\supset\mu$, and depends only on the skew diagram $\varkappa=\lambda\setminus\mu$. Moreover, it depends only on the shape of $\varkappa$, i.e. does not change if $\varkappa$ is shifted by a vector of the lattice $\mathbb{Z}\times\mathbb{Z}$. Thus, we associate to each skew diagram $\varkappa$ the positive element $\{\varkappa\}\in R_{|\varkappa|}$. We have:

$$t(\{\varkappa\}) = \{\varkappa^t\},$$

$$X^*(\{\varkappa\}) = \sum_{\varkappa'\perp\varkappa}\{\varkappa'\} \quad\text{and}\quad Y^*(\{\varkappa\}) = \sum_{\varkappa'\dashv\varkappa}\{\varkappa'\}.$$

Proof. By Theorem 4.3:

$$(X^*-1)(\{\lambda\setminus\mu\}) = (X^*-1)\circ\{\mu\}^*(\{\lambda\}) = \{\mu\}^*\circ(X^*-1)(\{\lambda\}) =$$

$$= \{\mu\}^*(\sum\{\lambda'\}) = \sum\{\lambda'\setminus\mu\},$$

where the sum is over all partitions λ' such that $\lambda'\perp\lambda, \lambda'\neq\lambda$ (similarly, for Y^*-1).

Using induction on $|\lambda|$, we can assume that our proposition is already known for all $\{\lambda'\setminus\mu\}$, where $\lambda'\perp\lambda$, $\lambda'\neq\lambda$. Therefore,

$(X^*-1)(\{\lambda\setminus\mu\}) = \sum\{\varkappa'\}$ (the sum is over skew diagrams $\varkappa'$ such that $\varkappa'\perp\varkappa, \varkappa'\neq\varkappa$; see 4.12).

The fact that $\{\lambda\setminus\mu\}$ depends only on the shape of $\varkappa$, follows at once from Proposition 4.4.

It remains to prove the equality $t(\{\varkappa\})=\{\varkappa^t\}$. This can be rewritten as

$$t(\{\lambda\setminus\mu\}) = \{\lambda^t\setminus\mu^t\}$$

hence it follows at once from 4.2 and 3.11 (*).

Q.E.D.

4.14. Let Y be a Young diagram and $\mathscr{X}$ a skew diagram. We compute the inner products

$$\langle x_Y, \{\mathscr{X}\}\rangle \text{ and } \langle y_Y, \{\mathscr{X}\}\rangle .$$

Once the answer is known ,one can compute the decomposition of $\{\mathscr{X}\}$(and so of any irreducible element in R) with respect to each of the bases $(x_Y^{\perp})$ and $(y_Y^{\perp})$ (see 3.17). Furthermore, if $\mathscr{X}=\lambda\setminus\mu$, where λ and μ are Young diagram, then by definition of $\mathscr{X}$ one has

$$x_Y \cdot \{\mu\} = \sum_{\lambda\in\mathcal{P}} \langle x_Y, \{\lambda\setminus\mu\}\rangle \cdot \{\lambda\} ;$$

$$y_Y \cdot \{\mu\} = \sum_{\lambda\in\mathcal{P}} \langle y_Y, \{\lambda\setminus\mu\}\rangle \cdot \{\lambda\} .$$

In particular, letting $\mu=\emptyset$, we obtain the decomposition of x_Y and y_Y with respect to Ω.

By a <u>numbering</u> of $\mathscr{X}$ we mean any mapping $\varphi:\mathscr{X}\longrightarrow \mathbb{N}$ which is a morphism of partially ordered sets where $\mathscr{X}$ is ordered by " $\leq_\rho$ " and $\mathbb{N}$ as usual (i.e. φ is such that $a,b\in\mathscr{X},\ a\leq_\rho b \Longrightarrow \varphi(a)\leq\varphi(b)$).

In terms of 4.12, $\varphi:\mathscr{X}\to\mathbb{N}$ is a numbering if and only if for any $k\in\mathbb{N}$ the set

$$\varphi^{-1}(\{1,2,\ldots,k\})$$

is a skew diagram, and $\varphi^{-1}(\{k\})$ its regular subset (the equivalence of these definitions is obvious). We say that φ has <u>type</u> $(n_1,n_2,\ldots)$ if $n_k = |\varphi^{-1}(k)|$ for $k\in\mathbb{N}$.

Geometrically, a numbering $\varphi:\mathscr{X}\to\mathbb{N}$ will be shown as an array of integers obtained by replacing each point x of

$\mathcal{X}$ by the number $\varphi(x)$. Such an array represents a numbering if and only if its numbers are non-decreasing along the rows and down the columns of $\mathcal{X}$; it has type $(n_1, n_2, \ldots)$ if any number $k \in \mathbb{N}$ occurs n_k times. We say that a numbering is row-strict (column-strict) if the numbers increase along the rows (down the columns) of $\mathcal{X}$.

Proposition. The inner product $\langle x_\nu, \{\mathcal{X}\}\rangle$ $(\langle y_\nu, \{\mathcal{X}\}\rangle)$ equals the number of column-strict (row-strict) numberings of $\mathcal{X}$ of type c.f. (ν) (see 3.7).

Proof. Let c.f. $(\nu) = (n_1, \ldots, n_r)$; we can assume that $|\mathcal{X}| = |\nu| = n_1 + \ldots + n_r$. We have

$$\langle x_\nu, \{\mathcal{X}\}\rangle = x_\nu^*(\{\mathcal{X}\}) = x_{n_1}^* \circ \ldots \circ x_{n_r}^*(\{\mathcal{X}\}).$$

By 4.13,

$$x_n^*(\{\mathcal{X}\}) = \sum \{\mathcal{X}'\}$$

(the sum is over skew diagrams $\mathcal{X}'$ such that $\mathcal{X}' \perp \mathcal{X}$ and $|\mathcal{X}'| = |\mathcal{X}| - n$). Applying several times this equality, we obtain that $\langle x_\nu, \{\mathcal{X}\}\rangle$ equals the number of chains

$$\emptyset = \mathcal{X}_0 \subset \mathcal{X}_1 \subset \mathcal{X}_2 \subset \ldots \subset \mathcal{X}_r = \mathcal{X},$$

where all $\mathcal{X}_i$ are skew diagrams, $\mathcal{X}_{k-1} \perp \mathcal{X}_k$, and $|\mathcal{X}_k| - |\mathcal{X}_{k-1}| = n_k$ for $k = 1, \ldots, r$. We associate with each such chain of diagrams the numbering $\varphi : \mathcal{X} \to \mathbb{N}$ which equals k on $\mathcal{X}_k \setminus \mathcal{X}_{k-1}$. The answer for $\langle x_\nu, \{\mathcal{X}\}\rangle$ follows; $\langle y_\nu, \{\mathcal{X}\}\rangle$ is computed in a similar way or by means of the automorphism t.

Q.E.D.

4.15. Remark. Proposition 4.14 has a combinatorial corollary: for any skew diagram $\mathcal{X}$ and the sequence $\alpha = (n_1, n_2 \ldots)$ of non-negative integers the number of column-strict numberings of $\mathscr{X}$ of type α does not depend on the order of numbers $n_1, n_2 \ldots$ (i.e. it depends only on $\mathscr{X}$ and the partition $\nu = (n_1, n_2 \ldots) \in \mathcal{P}$). Indeed, by 1.9 (c) this number equals $x^*_\nu(\{\mathscr{X}\})$.

4.16. Now we decompose the elements $\{\mathscr{X}\}$ (in particular, all irreducible elements of R) with respect to each of the bases (x_λ) and (y_λ) i.e. write $\{\mathscr{X}\}$ as polynomials in $x_1, x_2, \ldots$ and in $y_1, y_2, \ldots$.

Proposition. (a) Let $\lambda, \mu \in \mathcal{P}$, c.f.$(\lambda) = (l_1, \ldots, l_r)$ and c.f.$(\mu) = (m_1, \ldots, m_r)$ (see 3.7). Define the numbers $L_1 > L_2 > \ldots > L_r \geqslant 0$ and $M_1 > M_2 > \ldots > M_r \geqslant 0$ by

$$L_i = l_i + r - i, \quad M_i = m_i + r - i.$$

Then

$$\{\lambda \setminus \mu\} = \det(x_{L_i - M_j})_{i,j=1,\ldots,r} .$$

In particular, an irreducible element $\{\lambda\}$ equals

$$\det(x_{l_i - i + j})_{i,j=1,\ldots,r} .$$

(b) If c.f.$(\lambda^t) = (l'_1, \ldots, l'_s)$, c.f.$(\mu^t) = (m'_1, \ldots, m'_s)$, $L'_i = l'_i + s - i$ and $M'_i = m'_i + s - i$ then

$$\{\lambda \setminus \mu\} = \det(y_{L'_i - M'_j})_{i,j=1,\ldots,s} ;$$

in particular,

$$\{\lambda\} = \det(y_{l'_i - i + j})_{i,j=1,\ldots,s} ,$$

(recall that $x_k = y_k = 0$ for $k < 0$).

Proof. Put (temporarily)

$$[\lambda \setminus \mu] = \det(x_{L_i - M_j}),$$

It is easy to see that $[\lambda\setminus\mu]$ can be non-zero only if $L_i \geqslant M_i$ for all $i = 1,\dots,r$, i.e. only if $\lambda \supset \mu$; furthermore, $[\lambda\setminus\lambda] = 1 = \{\emptyset\}$. Using 4.4, 4.13 and induction on $|\lambda\setminus\mu|$, we see that the desired equality $\{\lambda\setminus\mu\} = [\lambda\setminus\mu]$ follows from the formula

$$(*) \qquad X^*([\lambda\setminus\mu]) = \sum_{\nu\perp\lambda} [\nu\setminus\mu]$$

To prove (*) we need some notation. If $\alpha_1,\dots,\alpha_r$ are rows of length r, we write $\Delta(\alpha_1;\alpha_2;\dots;\alpha_r)$ for the determinant of the matrix with these rows. For any $L\in\mathbb{Z}$ denote by $\alpha(L)$ the row $(x_{L-M_1}, x_{L-M_2},\dots,x_{L-M_r})$. Thus, by definition,

$$[\lambda\setminus\mu] = \Delta(\alpha(L_1);\alpha(L_2);\dots;\alpha(L_r)).$$

We apply the operator X^* to this determinant, using that X^* is a ring homomorphism, and $X^*(x_l) = \sum_{n\leqslant l} x_n$ (see 3.6, 3.1 (c)). We obtain:

$$X^*([\lambda\setminus\mu]) = \Delta\Big(\sum_{N_1\leqslant L_1}\alpha(N_1);\ \sum_{N_2\leqslant L_2}\alpha(N_2);\ \dots\ ;\ \sum_{N_r\leqslant L_r}\alpha(N_r)\Big)$$

Subtracting in this determinant the second row from the first one, then the third one from the second one etc., and then expanding it by polylinearity with respect to the rows, we see that

$$X^*([\lambda\setminus\mu]) = \sum \Delta(\alpha(N_1);\ \alpha(N_2);\dots;\ \alpha(N_r))$$

where the sum is over all $N_1,\dots,N_r$ such that

$$L_1 \geqslant N_1 > L_2 \geqslant N_2 > \dots > L_r \geqslant N_r .$$

In other words,

$$x^*([\lambda\setminus\mu]) = \sum [\nu\setminus\mu],$$

where the sum is over all partitions ν , whose canonical form c.f.$(\nu) = (n_1, n_2, \ldots, n_r)$ satisfies

(**) $\quad l_1 \geqslant n_1 \geqslant l_2 \geqslant n_2 \geqslant \ldots \geqslant l_r \geqslant n_r \geqslant 0$.

It remains to verify that (**) holds if and only if $\nu \perp \lambda$. But this is clear from the geometrical description of the relation „$\nu \perp \lambda$" (see 4.3). Indeed, the condition that ν is obtained by removing at most one point from each column of λ , means that all points removed from each row of λ lie to the right of the end of the next row.

This is just (**) !

(b) This follows from (a) by means of the automorphism t.

Q.E.D.

4.17. Now we shall compute the inner products $\langle z_\nu, \{\varkappa\}\rangle$ (we use the notation of 4.14). Taking into account 3.18, we see that this gives a solution of the following problems:

(a) Decomposition of the elements $\{\varkappa\}$ (in particular, of all irreducible elements $\{\lambda\}$) with respect to each of the bases (z_ν) and $(z_\nu^\perp)$:

$$\{\varkappa\} = \sum_{\nu\in\mathcal{P}} \frac{1}{c_\nu}\cdot\langle z_\nu, \{\varkappa\}\rangle\cdot z_\nu = \sum_{\nu\in\mathcal{P}} \langle z_\nu, \{\varkappa\}\rangle\cdot z_\nu^\perp .$$

(b) Decomposition of the products $z_\nu\cdot\{\mu\}$ (in particular, of the elements z_ν themselves) with respect to the basis Ω:

$$z_\nu\cdot\{\mu\} = \sum_{\lambda\in\mathcal{P}} \langle z_\nu, \{\lambda\setminus\mu\}\rangle\cdot\{\lambda\} .$$

By a skew-hook we shall mean any non-empty finite subset of $\mathbb{N}\times\mathbb{N}$ of the following form:

```
     xxxx
     x
     x
    xx
    x
xxxxx
```

more precisely, a <u>skew-hook</u> is a set $\{(i_1,j_1),(i_2,j_2),\ldots,(i_n,j_n)\}$, where for $k=1,\ldots,n-1$ the point (i_{k+1},j_{k+1}) is either $(i_k+1,\ j_k)$ or (i_k,j_k-1). Clearly, each skew-hook is a diagram. For any skew-hook $\varkappa$ put

$$s(\varkappa) = (-1)^{i-1} \quad ,$$

where i is the number of rows, intersecting with $\varkappa$. For example, the skew-hook displayed above meets 6 rows, so $s(\varkappa) = -1$.

Proposition. (a). We have

$$z_n^*(\{\varkappa\}) = \sum s(\varkappa')\cdot\{\varkappa\setminus\varkappa'\},$$

where the sum is over all skew-hooks $\varkappa'$ such that $|\varkappa'| = n$ and $\varkappa'$ is a regular subset of $\varkappa$ (see 4.12).

(b) The inner product $\langle z_\nu, \{\varkappa\}\rangle$ equals $\sum s(\varphi)$, where the sum is over all numberings $\varphi:\varkappa\to\mathbb{N}$ of type c.f.(ν) such that all subsets $\varphi^{-1}(k)$ for $k\in\mathbb{N}$ are skew-hooks, and

$$s(\varphi) = \prod_{k\in\mathbb{N}} s(\varphi^{-1}(k)).$$

Proof. Part (b) is derived from (a) as in the proof of Proposition 4.14. So we have only to prove (a).

Write $\varkappa$ as difference between two Young diagrams: $\varkappa = \lambda\setminus\mu$ (see 4.12). Let

$$L_1 > L_2 > \ldots > L_r \geq 0 \quad \text{and} \quad M_1 > M_2 > \ldots > M_r \geq 0$$

be coordinates of λ and μ, defined in 4.16. By 4.16 (a):

$$\{\varkappa\} = \det (x_{L_i - M_j})_{i,j=1,\ldots,r} \quad .$$

We apply the operator z_n^* to this determinant using that z_n^* is a derivation of R and $z_n^*(x_k) = x_{k-n}$ (see 3.17 (b)). We obtain:

$$(*) \qquad z_n^*(\{\varkappa\}) = \sum_{i_o=1}^{r} \Delta_{i_o}$$

where $\Delta_{i_o} = \det (x_{L_i - n\cdot\delta_{i,i_o} - M_j})$. We fix $1 \leq i_o \leq r$ and investigate Δ_{i_o}.

If $L_{i_o} - n = L_{i_1}$ for some $i_1 > i_o$ then Δ_{i_o} has two identical rows hence it equals 0. So we assume that

$$L_{i_1} > L_{i_o} - n > L_{i_1+1}$$

for some $i_1 \geqslant i_o$. Interchanging the i_o-th row of Δ_{i_o} successively with the (i_o+1)-th one, (i_o+2)-th one,..., i_1-th one, we see that

$$\Delta_{i_o} = (-1)^{i_1 - i_o} \cdot \{\lambda' \setminus \mu\},$$

where the "L-coordinates" of the partition λ' are

$$(L_1, L_2, \ldots, L_{i_o-1}, L_{i_o+1}, \ldots, L_{i_1}, L_{i_o}-n, L_{i_1+1}, \ldots, L_r).$$

Passing from these coordinates to usual ones, we see that if c.f.$(\lambda) = (l_1, .., l_r)$ then

$$\text{c.f.}(\lambda') = (l_1, \ldots, l_{i_o-1}, l_{i_o+1}-1, l_{i_o+2}-1, \ldots, l_{i_1}-1,$$
$$l_{i_o} - n + i_1 - i_o, l_{i_1+1}, \ldots, l_r).$$

Geometrically, this means that λ' is obtained from λ by removing the skew-hook $\varkappa'$ with $|\varkappa'| = n$ such that $\varkappa'$ meets the i-th row if and only if $i_0 \leq i \leq i_1$ (so $s(\varkappa') = (-1)^{i_1 - i_0}$). Clearly, the condition $L_{i_0} - n \neq L_{i_1}$ means that $\varkappa'$ is a regular subset of $\varkappa$, and, as i_0 varies from 1 to r, $\varkappa'$ runs over all such skew-hooks. Summarizing, we see that

$$\Delta_{i_0} = s(\varkappa') \cdot \{\varkappa \setminus \varkappa'\},$$

and (*) gives our assertion.

Q.E.D.

4.18. Now we decompose the elements $\{\varkappa\}$ with respect to the basis Ω , i.e. compute the inner products

$$\langle \{\nu\}, \{\varkappa\} \rangle,$$

where ν is a Young diagram and $\varkappa$ a skew diagram. Once the answer is known, we obtain the action of multiplication and comultiplication in the basis Ω , namely

$$\{\mu\} \cdot \{\nu\} = \sum_{\lambda \in \mathcal{P}} \langle \{\nu\}, \{\lambda \setminus \mu\} \rangle \cdot \{\lambda\} \qquad (\mu, \nu \in \mathcal{P});$$

$$m^*(\{\lambda\}) = \sum_{\mu, \nu \in \mathcal{P}} \langle \{\nu\}, \{\lambda \setminus \mu\} \rangle \cdot (\{\nu\} \otimes \{\mu\}) \qquad (\lambda \in \mathcal{P}).$$

(see 4.11).

Proposition. The inner product $\langle \{\nu\}, \{\varkappa\} \rangle$ equals the number of column-strict numberings φ of $\varkappa$ of type c.f.(ν) satisfying the following condition:

(J) If $\alpha_1, \alpha_2, \ldots, \alpha_n$ is the order of the numbers $\varphi(x)$ reading from right to left along the first row of $\varkappa$, next right to left along the second row, etc., then for any $m = 1, \ldots, n$ and $k \in \mathbb{N}$ the number of k's among $\alpha_1, \alpha_2, \ldots, \alpha_m$ is not less than the number of k+1's among $\alpha_1, \alpha_2, \ldots, \alpha_m$.

Example. $\nu = (3,2,1)$, $\varkappa = (3,3,1)\setminus(1)$. There are exactly two column-strict numberings of $\varkappa$ of type c.f.(ν), namely

$$\varphi_1 = \begin{matrix} & 1 & 1 \\ 1 & 2 & 2 \\ 3 & & \end{matrix} \qquad \text{and} \qquad \varphi_2 = \begin{matrix} & 1 & 1 \\ 1 & 2 & 3 \\ 2 & & \end{matrix}$$

The sequence $\alpha_1,\dots,\alpha_n$ for $\varphi=\varphi_1$ is 1,1,2,2,1,3, while for $\varphi=\varphi_2$ it is 1,1,3,2,1,2. Since in the latter sequence the number 3 precedes to all 2's φ_2 does not satisfy (J). Thus,

$$<\{\nu\},\{\varkappa\}> = 1$$

(by 4.14 we also obtain $<x_\nu, \{\varkappa\}> = 2$).

Proof. Denote the number of column-strict numberings of $\varkappa$ of type c.f.(ν) satisfying (J), by $g_{\varkappa,\nu}$. Clearly $g_{\varkappa,\nu} = 0$ unless $|\varkappa|=|\nu|$. We must prove that

$$\{\varkappa\} = \sum_{\bar{\nu}\in\mathcal{P}} g_{\varkappa,\bar{\nu}} \cdot \{\bar{\nu}\}.$$

This is evident for $|\varkappa| = 0$, while for $|\varkappa| > 0$ this follows from the equality

$$(*) \qquad (Y^* - 1)(\{\varkappa\}) = \sum_{\bar{\nu}\in\mathcal{P}} g_{\varkappa,\bar{\nu}} \cdot (Y^*-1)(\{\bar{\nu}\}).$$

(see 4.4). By 4.13 and induction on $|\varkappa|$, the left-hand side of (*) is

$$\sum_{\varkappa'} \{\varkappa'\} = \sum_{\varkappa'} \sum_{\nu\in\mathcal{P}} g_{\varkappa',\nu} \cdot \{\nu\},$$

where $\varkappa'$ runs over all skew diagrams such that $\varkappa' \neq \varkappa$ and $\varkappa' \dashv \varkappa$. Similarly, the right-hand side of (*) can be rewritten as

$$\sum_{\bar{\nu}\in\mathcal{P}} \sum_{\nu} g_{\varkappa,\bar{\nu}} \cdot \{\nu\},$$

where ν runs over all Young diagrams such that $\nu \dashv \bar{\nu}$ and $\nu \neq \bar{\nu}$. Comparing coefficients of each $\{\nu\}$, we see that (*) is equivalent to the following combinatorial statement:

for any skew diagram $\varkappa$ and **Young diagram** ν **with** $|\nu| < |\varkappa|$

$$(**) \quad \sum_{\varkappa' \dashv \varkappa} g_{\varkappa', \nu} = \sum_{\nu \dashv \bar{\nu}} g_{\varkappa, \bar{\nu}} .$$

This will be proved in Appendix 2.

Q.E.D.

4.19. We conclude this Chapter with a parametrization of irreducible elements in any PSH-algebra. By a <u>graded set</u> we shall mean a set X with a function

$$\deg : X \longrightarrow \mathbb{Z}^+ .$$

We denote by $S(X; \mathcal{P})$ the set of functions $\varphi : X \longrightarrow \mathcal{P}$ (see 3.7) such that $\varphi(x) = \emptyset$ but a finite number of $x \in X$. We consider $S(X; \mathcal{P})$ as a graded set by

$$\deg \varphi = \sum_{x \in X} \deg x \cdot |\varphi(x)|,$$

and put

$$S_n(X; \mathcal{P}) = \{ \varphi \in S(X; \mathcal{P}) \mid \deg \varphi = n \} .$$

Now let $\bar{R}$ be a PSH-algebra and $\mathcal{C}$ its set of irreducible primitive elements (we treat $\mathcal{C}$ as a graded set, the function deg being induced by the grading of $\bar{R}$). By 2.2 and 3.1,

$$\bar{R} = \bigotimes_{\rho \in \mathcal{C}} \bar{R}(\rho)$$

and each $\bar{R}(\rho)$ as a PSH-algebra is isomorphic up to grading to our universal PSH-algebra R. According to 3.1 (f), for any $\rho \in \mathcal{C}$ there are exactly two PSH-algebra isomorphisms $R \xrightarrow{\sim} \bar{R}(\rho)$;

let us choose one of them and denote it by φ_ρ.

Proposition. For any $\varphi \in S(\mathcal{C};\mathcal{P})$ put

$$\{\varphi\} = \prod_{\rho\in\mathcal{C}} \varphi_\rho(\{\varphi(\rho)\}) \in \bar{R} \quad \text{(see 4.1).}$$

The correspondence $\varphi \longmapsto \{\varphi\}$ is a bijection of $S(\mathcal{C};\mathcal{P})$ onto the (graded) set of irreducible elements of $\bar{R}$. In other words, modulo the choise of isomorphisms φ_ρ, irreducible elements of $\bar{R}_n$ are naturally parametrized by the set $S_n(\mathcal{C};\mathcal{P})$.

This follows at once from 2.2 and 4.1.

4.20. Remarks and complements. (a) The present proof of Theorem 4.3 is due to J.N.Bernstein. In the original author's proof the formula

$$(*) \qquad \{\lambda\} = \det(x_{l_i-i+j})$$

(see 4.16) was proved first; the computations in 4.16 show that (*) readily implies formulas 4.3. But formulas 4.3 look simple and natural, while the appearance of (*) in our approach seems rather mysterious, so we prefer the present proof.

Trying to understand (*), J.N.Bernstein has obtained the following beatiful formula. For any $l \geqslant 0$ define the operator $S_l: R \longrightarrow R$ by

$$S_l = \sum_{i\geqslant 0} (-1)^i x_{l+i} \circ y_i^*$$

(here x_{l+i} stands for the operator of multiplication by x_{l+i}). If $\lambda \in \mathcal{P}$, c.f.$(\lambda) = (l_1,\ldots,l_r)$ and $l \geqslant l_1$ then

$$(**) \qquad S_l(\{\lambda\}) = \{l, l_1, l_2, \ldots, l_r\}.$$

This implies that

$$\{\lambda\} = S_{l_1} \circ S_{l_2} \circ \ldots \circ S_{l_r}(1),$$

which is essentially equivalent to (*). The formula (**) can be easily derived from (*) by expanding the determinant expressing $\{l, l_1, \ldots, l_r\}$ up to first row. Since (**) will not be used in the sequel, we don't discuss it in more detail.

(b) In the original author's proof of Theorem 4.3 the important role was played by the following partial order relation on $\mathcal{P}$. By definition, $\mu < \lambda$ if μ can be obtained from λ by a chain of operations, each of which is a replacing of a pair $\{l, k\}$ of parts of a partition, where $l \geqslant k > 0$, by the pair $\{l+1, k-1\}$. This relation is well-known (see e.g. [10], [21]). The next statement shows its significance. For $\lambda, \mu \in \mathcal{P}$ the following three conditions are equivalent:

$$(1)\ \mu \leq \lambda \ ; \quad (2)\ x_\mu \leq x_\lambda \ ; \quad (3)\ \{\mu\} \leq x_\lambda .$$

This follows readily from the results of this section; we leave the proof to the reader (see [21]).

Chapter II. First applications

§ 5. Symmetric polynomials

In this section we realize the universal PSH-algebra R, as an algebra of symmetric polynomials in a countable set of indeterminates.

5.1. Let $\mathcal{H}$ be a commutative ring with unit. Consider the algebra $\mathcal{B}_{\mathcal{H}} = \mathcal{H}[[\xi_1, \xi_2, \dots]]$ of formal power series in a countable set of indeterminates $\xi_1, \xi_2, \dots$ over $\mathcal{H}$. Recall that $\mathcal{B}_{\mathcal{H}}$ consists of expressions of the form

$$\sum_{\alpha} h_\alpha \xi^\alpha ,$$

where α runs over all multiindices $(a_1, a_2, a_3, \dots)$ (all a_k are in $\mathbb{Z}^+$, and only finite number of them are non-zero), $\xi^\alpha = \xi_1^{a_1} \cdot \xi_2^{a_2} \dots$ and $h_\alpha \in \mathcal{H}$. The expressions ξ^α are called monomials, the number $|\alpha| = \sum a_k$ a degree of the monomial ξ^α. The group S_∞ of permutations of $\mathbb{N}$ acts on multiindices by

$$\sigma(a_1, a_2, \dots) = (a_{\sigma^{-1}(1)}, a_{\sigma^{-1}(2)}, \dots),$$

and on the algebra $\mathcal{B}_{\mathcal{H}}$ by

$$\sigma\left(\sum h_\alpha \xi^\alpha\right) = \sum h_\alpha \xi^{\sigma(\alpha)} .$$

Consider the subalgebra $A_{\mathcal{H}} \subset \mathcal{B}_{\mathcal{H}}$ consisting of power series $F \in \mathcal{B}_{\mathcal{H}}$, invariant under the action of S_∞, and such that degrees of all monomials occured in F, are uniformly bounded. Elements of $\mathcal{A}_{\mathcal{H}}$ are called symmetric polynomials over $\mathcal{H}$ in $\xi_1, \xi_2, \dots$. The most interesting for us will be the case when $\mathcal{H} = \mathbb{Z}$, so we shall write $\mathcal{A}$ for $\mathcal{A}_{\mathbb{Z}}$.

The algebra $\mathcal{A}_{\mathcal{H}}$ (in contrast to $\mathcal{B}_{\mathcal{H}}$) is graded by degrees of menomials; the correspondence $\mathcal{H} \mapsto \mathcal{A}_{\mathcal{H}}$ is a functor from the category of commutative rings to the category of commutative graded rings. Note that even in the case when $\mathcal{H}$ itself is graded, we forget this grading when considering the grading on $\mathcal{A}_{\mathcal{H}}$.

For any partition $\lambda \in \mathcal{P}$ we define the <u>symmetrized monomial</u> $u_\lambda \in \mathcal{A}_{\mathcal{H}}$ to be

$$u_\lambda = \sum \xi^\alpha ,$$

where the sum is over all multi-indices α conjugate to c.f.(λ) under the action of S_∞ (see 3.7). Evidently, $\mathcal{A}_{\mathcal{H}}$ is a free $\mathcal{H}$-module with the basis $\{u_\lambda \mid \lambda \in \mathcal{P}\}$.

It is also useful to realize $\mathcal{A}_{\mathcal{H}}$ as a projective limit of rings of symmetric polynomials in a finite number of indeterminates. More precisely, denote by

$$\mathcal{A}^N_{\mathcal{H}} = \bigoplus_{n \geq 0} (\mathcal{A}^N_{\mathcal{H}})_n$$

the graded algebra of symmetric polynomials over $\mathcal{H}$ in N indeterminates $\xi_1, \xi_2, \ldots, \xi_N$. There are natural morphisms

$$(\mathcal{A}_{\mathcal{H}})_n \longrightarrow (\mathcal{A}^N_{\mathcal{H}})_n \quad \text{and} \quad (\mathcal{A}^{N'}_{\mathcal{H}})_n \longrightarrow (\mathcal{A}^N_{\mathcal{H}})_n$$

for $N' > N$ ($\xi_i \longmapsto 0$ for $i > N$); denote any such morphism by φ_N. It is easy to see that these morphisms allow one to identify $(\mathcal{A}_{\mathcal{H}})_n$ with the projective limit

$$\varprojlim_{N \to \infty} (\mathcal{A}^N_{\mathcal{H}})_n .$$

Evidently the symmetrized monomials u_λ, where $\lambda \in \mathcal{P}_n$ and $r(\lambda) \leq N$ (see 3.7), form a $\mathcal{H}$-basis in $(\mathcal{A}^N_{\mathcal{H}})_n$. The mapping $\varphi_N : \mathcal{A}^{N'}_{\mathcal{H}} \longrightarrow \mathcal{A}^N_{\mathcal{H}}$ acts on u_λ by

$$\varphi_N(u_\lambda) = \begin{cases} u_\lambda & \text{if } r(\lambda) \leq N \\ 0 & \text{if } N < r(\lambda) \leq N' \end{cases} .$$

It follows that for $N \geq n$ the mappings

$$\varphi_N : (\mathcal{A}^{N'}_{\mathcal{H}})_n \longrightarrow (\mathcal{A}^{N}_{\mathcal{H}})_n \quad \text{and}$$

$$\varphi_N : (\mathcal{A}_{\mathcal{H}})_n \longrightarrow (\mathcal{A}^{N}_{\mathcal{H}})_n$$

are isomorphisms.

To avoid a confusion we shall sometimes writé

$$\mathcal{A}_{\mathcal{H}} = \mathcal{A}_{\mathcal{H}}(\xi_1, \xi_2, \ldots),$$

indicating explicitly the set of indeterminates. If $\Omega = \{\eta_1, \eta_2, \ldots\}$ is another countable set then any bijection $\varphi : \Omega \xrightarrow{\sim} \{\xi_1, \xi_2, \ldots\}$ induces an isomorphism

$$\mathcal{A}_{\mathcal{H}}(\Omega) \xrightarrow{\sim} \mathcal{A}_{\mathcal{H}}(\xi_1, \xi_2, \ldots) ;$$

clearly, this isomorphism does not depend on φ. We shall denote all such isomorphisms by C and identify canonically all $\mathcal{A}_{\mathcal{H}}(\Omega)$.

5.2. Now we construct an isomorphism of graded algebras $R \longrightarrow \mathcal{A} = \mathcal{A}_{\mathbb{Z}}$, where R is the PSH-algebra from §§ 3,4. The following construction is due to J.N.Bernstein.

(1) First, we construct the morphism of graded algebras

$$P : R \longrightarrow \mathcal{A}_R .$$

For any $N \geq 1$ define the mapping

$$P^{(N)} : R \longrightarrow R[\xi_1, \ldots, \xi_N]$$

to be the composition

$$R \xrightarrow{P_1} \underbrace{R \otimes R \otimes \ldots \otimes R}_{N \text{ times}} \xrightarrow{P_2} R[\xi_1] \otimes R[\xi_2] \otimes \ldots$$

$$\otimes R[\xi_N] \xrightarrow{P_3} R[\xi_1, \ldots, \xi_N],$$

where P_1, P_2, and P_3 are defined as follows.

The operator P_1 is obtained by iterating comultiplication in R, i.e. it equals the composition:

$$R \xrightarrow{m^*} R \otimes R \xrightarrow{m^* \otimes id} R \otimes R \otimes R \xrightarrow{m^* \otimes id} \dots \xrightarrow{m^* \otimes id} R \otimes R \otimes \dots \otimes R .$$

The operator P_2 is the tensor product of N copies of the operator $R \longrightarrow R[\xi]$, sending any $v \in R_n$ to $v \cdot \xi^n$. In other words, if $v_i \in R_{a_i}$ then

$$P_2(v_1 \otimes \dots \otimes v_N) = v_1 \xi_1^{a_1} \otimes v_2 \xi_2^{a_2} \otimes \dots \otimes v_N \xi_N^{a_N} ,$$

Finally, identifying

$$R[\xi_1, \dots, \xi_N] \text{ with } R[\xi_1] \otimes_R R[\xi_2] \otimes_R \dots \otimes_R R[\xi_N],$$

we define P_3 to be the natural projection. In other words, P_3 acts by

$$P_3(v_1 \xi_1^{a_1} \otimes v_2 \xi_2^{a_2} \otimes \dots \otimes v_N \xi_N^{a_N}) = v_1 v_2 \dots v_N \xi_1^{a_1} \xi_2^{a_2} \dots \xi_N^{a_N}.$$

Clearly, P_1, P_2, P_3, and so $P^{(N)}$ are homomorphisms of graded rings (for P_1 this follows from the Hopf axiom (H) from 1.3). Commutativity and associativity of comultiplication in R imply that

$$P^{(N)}(R) \subset \mathcal{A}_R^N$$

i.e. that $P^{(N)}$ is a morphism of graded rings $R \longrightarrow \mathcal{A}_R^N$. Evidently,

$$\varphi_N \circ P^{(N')} = P^{(N)} \qquad \text{for} \quad N' > N ;$$

tending N to infinity, we obtain the desired homomorphism

$$P : R \longrightarrow \mathcal{A}_R .$$

(2) Consider the ring homomorphism $\delta_x : R \longrightarrow \mathbb{Z}$ (see 3.6). Since the correspondence $\mathcal{H} \longmapsto \mathcal{A}_{\mathcal{H}}$ is a functor, δ_x induces the graded ring morphism

$$\mathcal{A}_R \longrightarrow \mathcal{A}_{\mathbb{Z}} = \mathcal{A},$$

which will be denoted also by δ_x. We define the graded ring morphism $p:R \longrightarrow \mathcal{A}$ to be the composition

$$R \xrightarrow{P} \mathcal{A}_R \xrightarrow{\delta_x} \mathcal{A} .$$

5.3. Proposition. (a) The morphism p acts on various elements of R in accordance to the following table.

$v \in R$	$p(v) \in \mathcal{A}$
x_n (see 3.1)	$\sum_{\lvert\lambda\rvert = n} u_\lambda$ (the sum of all monomials of degree n)
y_n (see 3.1)	$u_{(1^n)}$ (elementary symmetric polynomial)
z_n (see 3.15)	$u_{(n)}$ (power sum)
$x_\lambda^\perp$ (see 3.17)	u_λ
$\{\mathfrak{X}\}$ (see 4.12, 4.13)	$\sum_\varphi \prod_{x \in \mathfrak{X}} \xi_{\varphi(x)}$ (φ runs over all column-strict numberings of the skew diagram $\mathfrak{X}$; see 4.14)

(b) The morphism $p:R \longrightarrow \mathcal{A}$ is an isomorphism. We shall identify R and $\mathcal{A}$ via this isomorphism and use the same notation for an element of R and its image in $\mathcal{A}$.

<u>Proof</u>. Let $v \in R_n$. Using the definition of δ_x and the fact that δ_x is multiplicative (see 3.6) we can readily compute

$$\varphi_N \circ p(v) = \delta_x \circ P_3 \circ P_2 \circ P_1(v) \qquad \text{(see 5.1, 5.2)}$$

We obtain that

$$(*) \quad p(v) = \sum_{\lambda \in \mathcal{P}_n} x_\lambda^*(v) \cdot u_\lambda = \sum_{\lambda \in \mathcal{P}_n} \langle x_\lambda, v \rangle \cdot u_\lambda .$$

All formulas of the part (a) follow at once from (*) and the results of §§ 3,4 (see 3.17, 4.14, and 4.15). Part (b) follows at once from the equality $p(x_\lambda^\perp) = u_\lambda$ (see 3.17).

Q.E.D.

5.4. The isomorphism $p: R \longrightarrow \mathcal{A}$ makes $\mathcal{A}$ into a PSH-algebra. All results of §§ 3,4 are transformed to the assertions on symmetric polynomials. All these assertions are classical. We will not formulate them since this can be done automatically; let us only mention some names.

The symmetric polynomials $\{\lambda\}$ where $\lambda \in \mathcal{P}$, are called <u>Schur functions</u> or <u>S-functions</u>, while the $\{\mathcal{X}\}$, where $\mathcal{X}$ is a skew diagram, are called skew Schur functions. The formulas 3.16 are <u>Newton formulas</u>, the expressions 4.16 (a) and (b) of S- functions are the <u>Jacobi-Trudi</u> and the <u>Naegelsbach-Kostka</u> formulas, while the identity

$$\det (x_{L_i - M_j}) = \det (y_{L_i' - M_j'})$$

is the <u>Aitken theorem</u>. Proposition 4.18 gives the <u>Littlewood-</u>

Richardson rule for multiplying S- functions, and the Littlewood-Roe theorem, which says that the coefficient of $\{\nu\}$ in $\{\lambda\setminus\mu\}$ (when written as a linear combination of Schur functions) is equal to the coefficient of $\{\lambda\}$ in $\{\mu\}\cdot\{\nu\}$.

The classical proofs of these statements can be found in [22], Ch. VI (two another approaches are developed in [10] and [23]). Our proofs, which are based on the systematic use of the "lowering" operators a^*, sometimes seem to be more simple and natural (especially, those of the Littlewood-Richardson rule and the Littlewood-Roe theorem). The lowering operators in $\mathcal{A}$ were used by a number of authors e.g. by MacMahon, Van der Corput and Foulkes; the references and the discussion of them from the same point of view of Hopf algebra can be found in [5].

5.5. In the remainder of this section we shall express the various structures on $\mathcal{A}$ arising from the identification of $\mathcal{A}$ and R, in terms of the "coordinates" ξ_i i.e. using the embedding of $\mathcal{A}$ into the power series ring

$$\mathcal{B}=\mathbb{Z}[[\xi_1,\xi_2,\dots]].$$

First we want to write down explicitly the comultiplication in $\mathcal{A}$. We introduce two countable sets of indeterminates $\eta = \{\eta_1,\eta_2,\dots\}$ and $\zeta=\{\zeta_1,\zeta_2,\dots\}$. By means of the mapping

$$F\otimes G\longmapsto F(\eta)\cdot G(\zeta)$$

we realize $\mathcal{A}\otimes\mathcal{A}$ as the subalgebra in $\mathbb{Z}[[\eta,\zeta]]$ consisting of those series which are symmetric polynomials separately with respect to η and ζ.

Proposition. If $\mathcal{A}\otimes\mathcal{A}$ is realized as above, the comultiplication $m^*: \mathcal{A} \longrightarrow \mathcal{A}\otimes\mathcal{A}$ has the form

$$\mathcal{A} \xrightarrow{C} \mathcal{A}(\eta,\zeta) \hookrightarrow \mathcal{A}\otimes\mathcal{A},$$

where C is the canonical isomorphism (see 5.1).

Proof. It suffices to verify the equality $m^*(x)=C(x)$ when x varies over some set of generators of the ring $\mathcal{A}$. We choose as generators the power sums

$$z_n = \sum_i \xi_i^n \qquad (n = 1,2,\dots)$$

(they are generators of $\mathcal{A}_{\mathbb{Q}}$ only, but this does not matter). By 5.3 and 3.15 $m^*(z_n) = z_n \otimes 1 + 1 \otimes z_n$. On the other hand, by definition of C we have

$$C(z_n) = \sum_i \eta_i^n + \sum_i \zeta_i^n = z_n \otimes 1 + 1 \otimes z_n.$$

Q.E.D.

5.6. Now we discuss the inner product $\langle\,,\rangle$ in $\mathcal{A}$. We recall that it is induced by the T-group structure on $\mathcal{A}$ (see 1.2); by definition, the S-functions $\{\lambda\}$ $(\lambda \in \mathcal{P})$ form an orthonormal basis in $\mathcal{A}$. We want to describe $\langle\,,\rangle$ more explicitly in a "coordinate" form.

It is known from standard linear algebra that an inner product on a finite dimensional vector space V can be characterized by a non-degenerate symmetric tensor $g \in V \otimes V$. Namely, if $\langle\,,\rangle$ is an inner product on V, $\{e_1,\dots,e_p\}$ is any basis of V, and $\{e_1^\perp,\dots,e_p^\perp\}$ is the dual basis of V with respect to $\langle\,,\rangle$ then

$$g = \sum_{i=1}^{p} e_i \otimes e_i^\perp \in V \otimes V$$

It is easy to see that g depends only on $\langle , \rangle$. Conversely, $\langle , \rangle$ is determined by g; in fact, g determines naturally the inner product on the dual space V^*, this allows one to identify V and V^*, and hence to obtain the inner product on V.

We will describe explicitly the tensor $g_n \in \mathcal{A}_n \otimes \mathcal{A}_n$ corresponding to the inner product $\langle , \rangle$ on $\mathcal{A}_n$.

<u>Proposition</u>. If $\mathcal{A} \otimes \mathcal{A}$ is realized as in 5.5 then g_n is equal to the sum of all monomials of degree n in the set of indeterminates

$$\eta \cdot \zeta = \{ \eta_i \cdot \zeta_j \quad | \quad i,j = 1,2,\dots \} .$$

<u>Proof</u>. By 5.3 it suffices to verify the identity

$$(*) \qquad \sum_{|\lambda|=n} x_\lambda (\eta) \cdot u_\lambda (\zeta) = x_n(\eta \cdot \zeta).$$

One can do it by a straightforward computation (see [10], pp. 37-38), but we prefer another approach based on the homomorphism

$$P : R \longrightarrow \mathcal{A}_R$$

from 5.2 (1). We identify $\mathcal{A}_R$ with $R \otimes \mathcal{A}$ via

$$\sum_\lambda v_\lambda \cdot u_\lambda \longmapsto \sum_\lambda v_\lambda \otimes u_\lambda \qquad (v_\lambda \in R),$$

so P becomes a homomorphism from R to $R \otimes \mathcal{A}$. Identifying R with $\mathcal{A}$ by means of the isomorphism p, we obtain the homomorphism

$$P : \mathcal{A} \longrightarrow \mathcal{A} \otimes \mathcal{A} .$$

By definition of P and 3.1 (d),

$$P(x_n) = \sum_{|\lambda|=n} x_\lambda \otimes u_\lambda$$

Therefore the identity (*) follows from the next

5.7. Lemma. If $\mathcal{A}\otimes\mathcal{A}$ is realized as in 5.5, the homomorphism $P : \mathcal{A} \longrightarrow \mathcal{A}\otimes\mathcal{A}$ acts by

$$P(F)(\eta,\zeta) = F(\eta\cdot\zeta).$$

Proof. As in 5.5, we see that it suffices to verify our identity only for $F = z_n$ $(n = 1,2,\dots)$. By definition of P, primitivity of z_n, and 5.3 we see that

$$P(z_n) = z_n \otimes u_{(n)} = z_n \otimes z_n.$$

Hence,

$$P(z_n) = \Big(\sum_i \eta_i^n\Big)\cdot\Big(\sum_j \zeta_j^n\Big) = \sum_{i,j} (\eta_i\cdot\zeta_j)^n = z_n(\eta\cdot\zeta)$$

Q.E.D.

5.8. Now we consider the graded group

$$\mathcal{A}^* = \bigoplus_{n\geq 0} \mathcal{A}_n^*,$$

where $\mathcal{A}_n^*$ is dual of $\mathcal{A}_n$. Since all $\mathcal{A}_n$ are finite dimensional, the operator adjoint to multiplication (comultiplication) in $\mathcal{A}$ induces comultiplication (multiplication) on $\mathcal{A}^*$, i.e. $\mathcal{A}^*$ is also a Hopf algebra. The inner product $\langle\ ,\ \rangle$ on $\mathcal{A}$ gives rise to the isomorphism

$$g : \mathcal{A} \xrightarrow{\sim} \mathcal{A}^*;$$

the axiom (S) from 1.4 means that g is an isomorphism of Hopf algebras. We want to express these structures on $\mathcal{A}^*$ in a "coordinate" form.

Let $\mathcal{B}^*$ be the algebra of differential operators on $\mathcal{B}_{\mathbb{Q}}$ (see 5.1) i.e. the algebra over $\mathbb{Q}$, generated by the operators

$$\frac{\partial}{\partial \xi_i} : \mathcal{B}_{\mathbb{Q}} \longrightarrow \mathcal{B}_{\mathbb{Q}} ;$$

we define the grading on $\mathcal{B}^*$ by $\deg \frac{\partial}{\partial \xi_i} = 1$. The group S_∞ acts on $\mathcal{B}^*$ by

$$\sigma(D) = \sigma \circ D \circ \sigma^{-1} \qquad \text{(see 5.1)};$$

clearly,

$$\sigma\left(\frac{\partial}{\partial \xi_i}\right) = \frac{\partial}{\partial \xi_{\sigma(i)}} .$$

Denote by $\mathcal{B}^*/S$ the quotient of $\mathcal{B}^*$ by the subspace spanned by operators of the form

$$\sigma D - D \qquad (\sigma \in S_\infty ,\ D \in \mathcal{B}^*).$$

Clearly, $\mathcal{B}^*/S = \bigoplus_{n \geq 0} (\mathcal{B}^*/S)_n$ is a graded vector space over $\mathbb{Q}$; note that $\mathcal{B}^*/S$ is <u>not</u> an algebra under the operation of composition of operators. We will use the same notation for an element of $\mathcal{B}^*$ and its image in $\mathcal{B}^*/S$.

For any multiindex $\alpha = (a_1, a_2, \dots)$ put

$$D_\alpha = \frac{1}{a_1!\, a_2! \dots} \left(\frac{\partial}{\partial \xi_1}\right)^{a_1} \circ \left(\frac{\partial}{\partial \xi_2}\right)^{a_2} \circ \dots ;$$

if $\lambda \in \mathcal{P}$ then we write D_λ instead of $D_{c.f.(\lambda)}$ (see 3.7). Clearly, the elements D_λ ($\lambda \in \mathcal{P}_n$) form a basis in $(\mathcal{B}^*/S)_n$. With each element $D \in (\mathcal{B}^*/S)_n$ there is associated the functional $(\mathcal{A}_{\mathbb{Q}})_n \longrightarrow \mathbb{Q}$, which will be denoted also by D. It is easy to see that

$$D_\mu(u_\lambda) = \delta_{\lambda\mu}$$

for $\lambda, \mu \in \mathcal{P}_n$. This proves the following.

Proposition. The graded group $\mathcal{A}^*$ is naturally identified with the lattice in $\mathcal{B}^*/S$, spanned by the elements D_λ $(\lambda \in \mathcal{P})$. The isomorphism $g : \mathcal{A} \xrightarrow{\sim} \mathcal{A}^*$ (see above) acts by

$$g(x_\lambda) = D_\lambda \qquad (\lambda \in \mathcal{P}).$$

5.9. We describe explicitly the multiplication and comultiplication in $\mathcal{A}^*$, where $\mathcal{A}^*$ is realized as in 5.8. As in 5.5, let us introduce two sets of indeterminates η and ζ . We identify $\mathcal{B}^* \otimes \mathcal{B}^*$ with $\mathcal{B}^*(\eta,\zeta)$ via the mapping:

$$D_1 \otimes D_2 \longmapsto D_1(\eta) \circ D_2(\zeta).$$

Therefore, $\mathcal{A}^* \otimes \mathcal{A}^*$ is identified with the lattice in $\mathcal{B}^*(\eta,\zeta)/S_\infty(\eta) \times S_\infty(\zeta)$, spanned by the operators $D_\lambda(\eta) \circ D_\mu(\zeta)$ $(\lambda, \mu \in \mathcal{P})$.

Proposition. (a) The multiplication $\mathcal{A}^* \otimes \mathcal{A}^* \longrightarrow \mathcal{A}^*$ is induced by an isomorphism $\mathcal{B}^*(\eta,\zeta) \xrightarrow{\sim} \mathcal{B}^*$ obtained by means of an (arbitrary) bijection $\{\eta_1, \eta_2, \ldots, \zeta_1, \zeta_2, \ldots\} \xrightarrow{\sim} \{\xi_1, \xi_2, \ldots\}$.

(b) The comultiplication $\mathcal{A}^* \longrightarrow \mathcal{A}^* \otimes \mathcal{A}^*$ is induced by the ring homomorphism $\mathcal{B}^* \longrightarrow \mathcal{B}^*(\eta,\zeta)$ (with respect to composition as a multiplication), sending $\frac{\partial}{\partial \xi_i}$ to $\frac{\partial}{\partial \eta_i} + \frac{\partial}{\partial \zeta_i}$ $(i=1,2,\ldots)$.

To prove this proposition it suffices to apply Proposition 5.5, taking into account all definitions and identifications of 5.5 and 5.8; we leave this to the reader.

5.10. Now we describe the operators v^* on $\mathcal{A}$ in terms of differential operators.

Proposition. Let $v \in \mathcal{A}$ and $g(v) = D \in \mathcal{A}^*$ (see 5.8).

Then for all $F \in \mathcal{A}(\xi)$

$$v^*(F)\,(\xi) = D(\eta)\,\left[F(\xi,\eta)\right]\Big|_{\eta=0}$$

(this means that we must write down the differential operator D in indeterminates η, evaluate it at the polynomial $F(\xi,\eta)$ and then put $\eta_i = 0$ $(i = 1,2,\ldots)$.

This follows at once from 1.9 (a),(b), 5.5 and 5.8.

5.11. Corollary. The operator $X^*: \mathcal{A} \longrightarrow \mathcal{A}$ (see 3.6) equals the composition

$$\mathcal{A}(\xi_1,\xi_2,\ldots) \xrightarrow{\;C\;} \mathcal{A}(\xi_0,\xi_1,\xi_2,\ldots) \xrightarrow{\;\text{"}\xi_0 \mapsto 1\text{"}\;} \mathcal{A}(\xi_1,\xi_2,\ldots),$$

where C is the canonical isomorphism from 5.1, and the map "$\xi_0 \mapsto 1$" takes $F(\xi_0,\xi_1,\xi_2,\ldots)$ to $F(1,\xi_1,\xi_2,\ldots)$.

Proof. By 5.8, $g(x_n) = \frac{1}{n!}\left(\frac{\partial}{\partial \xi_0}\right)^n$. Hence our assertion follows from 5.10.

Q.E.D.

5.12. We conclude this section with a classical expression of S - functions $\{\lambda\}$ in terms of alternating polynomials. Fix $N \geqslant 1$ and put

$$\Delta(\xi_1,\ldots,\xi_N) = \prod_{1\leq i<j\leq N} (\xi_i - \xi_j) .$$

Clearly, Δ is an alternating polynomial in $\xi_1,\ldots,\xi_N$, and the operator of multiplication by Δ is an isomorphism

$$\mathcal{A}_n^N \longrightarrow \Lambda_n^N ,$$

where Λ_n^N is the group of all integral alternating polynomials in $\xi_1,\ldots,\xi_N$ of degree $n + C_N^2$. We assign to each partition

$\lambda \in \mathcal{P}_n$ with $r(\lambda) \leq N$ and c.f.$(\lambda) = (l_1,\dots,l_N)$ (see 3.7) the polynomial $\Delta_\lambda(\xi_1,\dots,\xi_N) \in \mathcal{A}_n^N$ obtained by alternating the monomial

$$\xi_1^{l_1+N-1}\ \xi_2^{l_2+N-2} \cdot \dots \cdot \xi_N^{l_N} \ ;$$

in other words,

$$\Delta_\lambda(\xi_1,\dots,\xi_N) = \det(\xi_i^{L_j})_{i,j=1,\dots,N} \ ,$$

where $L_1 > L_2 > \dots > L_N \geq 0$ are the coordinates of λ described in 4.16. Evidently, the elements Δ_λ, where $\lambda \in \mathcal{P}_n$, $r(\lambda) \leq N$, form a $\mathbb{Z}$-basis in $\mathcal{A}_n^N$, so the fractions Δ_λ / Δ form a $\mathbb{Z}$-basis in $\mathcal{A}_n^N$. If $r(\lambda) > N$, we put $\Delta_\lambda(\xi_1,\dots,\xi_N) = 0$

<u>Proposition</u>. For $N \geq 1$ and $\lambda \in \mathcal{P}$ we have

$$\varphi_N(\{\lambda\}) = \Delta_\lambda(\xi_1,\dots,\xi_N)/\Delta(\xi_1,\dots,\xi_N) \text{ (see 5.1,5.3)} .$$

<u>Proof</u>. Denote temporarily the fraction

$$\Delta_\lambda(\xi_1,\dots,\xi_N)/\Delta(\xi_1,\dots,\xi_N) \in \mathcal{A}_{|\lambda|}^N$$

by $[\lambda]_N$. One can verify directly that

$$\varphi_N([\lambda]_{N'}) = [\lambda]_N \qquad \text{for } N' > N \ .$$

Passing to projective limit, we obtain the element $[\lambda] \in \mathcal{A}_{|\lambda|}$ such that $\varphi_N([\lambda]) = [\lambda]_N$. We must prove that $[\lambda] = \{\lambda\}$. According to the assertion (*) from 4.4 it suffices to verify that $[\emptyset] = 1$, and

$$(*) \qquad X^*[\lambda] = \sum_{\mu \perp \lambda} [\mu] \ .$$

The identity $[\emptyset] = 1$ means that $\Delta_\emptyset = \Delta$; this is a well-known formula for the Vandermonde determinant. To prove (*) we fix $N \geq |\lambda|$. Since

$$\varphi_N : \mathcal{A}_{|\lambda|} \longrightarrow \mathcal{A}^N_{|\lambda|}$$

is an isomorphism (see 5.I), it suffices to verify (*) in $\mathcal{A}^N_{|\lambda|}$. By 5.II, we have

$$\varphi_N(x^*[\lambda])(\xi_1,\dots,\xi_N) = [\lambda]_{N+1}(\xi_1,\xi_2,\dots,\xi_N,1).$$

Therefore, we have only to verify the identity

$$[\lambda]_{N+1}(\xi_1,\dots,\xi_N,1) = \sum_{\mu\perp\lambda}[\mu]_N .$$

This can be done by an easy computation entirely similar to that in 4.I6; the details are left to the reader.

Q.E.D.

5.I3. Remarks. (a) Clearly, the construction of the homomorphism $P : \mathcal{H} \longrightarrow \mathcal{A}_{\mathcal{H}}$ from 5.2 (I) makes sense for any Hopf algebra $\mathcal{H}$ with commutative multiplication and comultiplication. Its meaning is not yet well understood.

(b) The most natural approach to the investigation of the algebra $\mathcal{A}$ in the spirit of this work would be to introduce in $\mathcal{A}$ independently the PSH-algebra structure and apply the structural theory of Chapter I. The Hopf algebra structure on $\mathcal{A}$ is quite natural (see 5.5). The main difficulty is to introduce the T-group structure on $\mathcal{A}$ and to verify axioms (P) and (S) (see I.2, I.4). Proposition 5.I2 shows that the T-group structure on $\mathcal{A}$ becomes more natural when passing from symmetric polynomials to alternating ones, but the meaning of the axiom (P) in this realization is not clear.

(c) The realization of R as a ring of symmetric polynomials has usual advantages and drawbacks of a coordinate approach. The advantage is that it makes some computations easier, while the main drawback is that some symmetries break down. For example, passing to the coordinates ξ_i , we loose the symmetry between the elements x_n and y_n, which is evident from Theorem 3.I. We don't know a good expression for the automorphism $t : \mathcal{A} \longrightarrow \mathcal{A}$ or the operator $Y^* : \mathcal{A} \longrightarrow \mathcal{A}$ in terms of the coordinates (ξ_i).

§ 6. Representations of symmetric groups

In this section we apply results of §§ 3,4 to the representation theory of the symmetric groups S_n .

6.I. First, we introduce general terminology relative to representations of finite groups. We write $\mathcal{A}(G)$ for the category of finite-dimensional complex representations of a finite group G, and R(G) for its Grothendieck group. R(G) is a T-group (see I.2) with the set $\Omega = \Omega(G)$ consisting of the equivalence classes of irreducible representations of G. Evidently, positive elements of R(G) are identified with the isomorphism classes of objects of $\mathcal{A}(G)$; we shall use the same notation for a representation of G and its image in R(G). By the Schur lemma

$$\langle \pi, \tau \rangle = \dim \operatorname{Hom}(\pi, \tau) \quad \text{for} \quad \pi, \tau \in \mathcal{A}(G) .$$

Any additive functor $\Phi : \mathcal{A}(G) \longrightarrow \mathcal{A}(H)$ (this means that Φ sends direct sums to direct sums) induces the positive morphism $R(G) \longrightarrow R(H)$, which will be denoted also by Φ. In particular, if H is a subgroup in G then there are defined the homomorphisms:

$$\operatorname{Ind}_H^G : R(H) \longrightarrow R(G) \quad \text{and} \quad \operatorname{Res}^G_H : R(G) \longrightarrow R(H)$$

(induction and restriction); by the Frobenius reciprocity they are ajoint to each other with respect to inner products $\langle \, , \rangle$.

The tensor product of representations gives a bifunctor

$$\otimes : \mathcal{A}(G) \times \mathcal{A}(H) \longrightarrow \mathcal{A}(G \times H),$$

additive with respect to each argument. It induces the T-group morphism

$$R(G) \otimes R(H) \longrightarrow R(G \times H),$$

which is known to be an isomorphism. We identify $R(G \times H)$ with $R(G) \otimes R(H)$ via this isomorphism.

Denote by $C(G)$ the complex vector space of class functions on G. Assigning to any representation of G its character, we identify $R(G)$ with the lattice in $C(G)$; we will denote identically an element of $R(G)$ and its image in $C(G)$. We extend the inner product $\langle , \rangle$ on $R(G)$ to the Hermitian inner product on $C(G)$, given by

$$\langle f_1, f_2 \rangle = \frac{1}{|G|} \sum_{g \in G} f_1(g) \cdot \overline{f_2(g)} .$$

We extend any $\mathbb{Z}$-linear map $\varphi : R(G) \longrightarrow R(H)$ to the $\mathbb{C}$-linear map $C(G) \longrightarrow C(H)$, which will be denoted also by φ.

6.2. Set $R(S) = \bigoplus_{n \geq 0} R(S_n)$. The group $R(S)$ is a graded T-group. We define the multiplication and comultiplication in $R(S)$ as indicated above in I.1.

<u>Proposition</u>. $R(S)$ is a PSH-algebra. It has a unique irreducible primitive element, namely the identity representation of the group $S_1 = \{e\}$.

<u>Proof</u>. The axioms (G), (U), (U^*), (Con), (T), and (P) (see I.3, I.4) are satisfied evidently. The axiom (S) follows from the Frobenius reciprocity. The axiom (H) follows from the Mackey theorem; for details see Appendix 3, A3.2. Finally, it was shown in I.6 that the axioms (A), (A^*), (Com) and (Com^*) follow from other ones. Thus, $R(S)$ is a PSH-algebra.

By definition, an irreducible primitive element in $R(S)_n$ is an irreducible representation of S_n, whose restriction to any subgroup $S_k \times S_l$ where $k,l > 0$, $k+l = n$, is zero. Evidently this can be only when $n=1$. Proposition follows.

Q.E.D.

6.3. By 3.1 (g), R(S) is isomorphic as a PSH- algebra to the algebra R from §§ 3,4. There are exactly two PSH-algebra isomorphisms $R \xrightarrow{\sim} R(S)$ (see 3.1 (g)). We choose the one sending the element x_2 to the identity representation of S_2 (by 3.1 (f) it is uniquely determined by this property) and identify R(S) with R via this isomorphism.

Now we give a dictionary which shows the meaning in R(S) of certain elements, operators and constants relative to R. We need some notation. Write ε for the sign character of any group S_n. For any ordered partition $\alpha = (a_1, \ldots, a_r)$ of n denote by S_α the group $S_{a_1} \times S_{a_2} \times \ldots \times S_{a_r}$; it naturally embeds in S_n, so we consider S_α as a subgroup in S_n. For $0 \leq k \leq n$ denote by S_k' the subgroup of S_n, consisting of permutations which are identity on the subset $\{1,2,\ldots,n-k\}$. It is well-known that conjugacy classes in S_n are parametrized by partitions of n: to a partition $\lambda = (l_1,\ldots,l_r) \in \mathcal{P}_n$ there corresponds the class C_λ consisting of permutations with cycle lengths $l_1,l_2,\ldots,l_r$. Denote by $\chi_\lambda \in C(S_{|\lambda|})$ the characteristic function of C_λ.

<u>Proposition</u>. The various notions relative to R are translated to the language of R(S) in accordance with the following table

	R	R(S)
1.	ρ^n	The regular representation of S_n
2.	$\rho^*: R_n \to R_{n-1}$ (see 1.9 (b))	The restriction from S_n to S_{n-1}

R	R (S)
3. x_n (see 3.I(b))	The identity representation of S_n
4. y_n (see 3.I (b))	The character ε of S_n
5. t (see 3.I (e), 3.II)	The (inner) tensor product by ε
6. x_k^* (see I.9 (b))	If $\pi \in \mathcal{A}(S_n)$ acts on a space V then $x_k^*(\pi)$ is the representation of S_{n-k} in the subspace of S_k' - invariant vectors in V
7. x_λ (see 3.8)	$\mathrm{Ind}_{S_\alpha}^{S_{\lvert\lambda\rvert}} 1$, where α is any ordering of parts of λ
8. y_λ (see 3.8)	$\mathrm{Ind}_{S_\alpha}^{S_{\lvert\lambda\rvert}} \varepsilon$ (α as in the preceding formula)
9. $z_\lambda^\perp$ (see 3.I7)	χ_λ
I0. c_λ (see 3.I7(c))	$\lvert S_{\lvert\lambda\rvert}\rvert / \lvert C_\lambda \rvert$
II. $\{\lambda\}$ (see 4.I)	The irreducible representations of the groups S_n

Proof. The assertions 1 and 11 are trivial, while 2-8 follow at once from definitions and the items indicated in the first column of the table. It remains to prove assertions 9 and I0.

The crucial point is that $\chi_{(n)}$ is a primitive element of the Hopf algebra

$$C(S) = \bigoplus_{n \geq 0} C(S_n) .$$

This is obvious, since no cycle of length n can be contained in the subgroup $S_{(k,l)} \subset S_n$ for $k,l > 0$, $k+l = n$.

By 3.15 (a), z_n is proportional to $\chi_{(n)}$. Evidently,

$$\langle x_n, \chi_{(n)} \rangle = |C_{(n)}| / |S_n| ,$$

so
$$z_n = |S_n| / |C_{(n)}| \cdot \chi_{(n)} .$$

Now we apply the formula for the action of induction on class functions (see [9], 7.2 or 8.2 below): if H is a subgroup of a finite group G, C is a conjugacy class in H, and D is the conjugacy class in G containing C, then

$$(*) \qquad \mathrm{Ind}_H^G \left({}^{|H|}/_{|C|} \cdot \chi_C \right) = {}^{|G|}/_{|D|} \cdot \chi_D .$$

Let $\lambda = (1_1, \ldots, 1_r) \in \mathcal{P}_n$. Applying $(*)$ to the case when $G = S_n$, $H = S_{1_1} \times \ldots \times S_{1_r}$, $C = C_{(1_1)} \times \ldots \times C_{(1_r)}$, and $D = C_\lambda$, we obtain:

$$z_\lambda = {}^{|S_n|}/_{|C_\lambda|} \cdot \chi_\lambda$$

It follows that

$$c_\lambda = \langle z_\lambda , z_\lambda \rangle = (|S_n|/|C_\lambda|)^2 \cdot \langle \chi_\lambda , \chi_\lambda \rangle = |S_n|/|C_\lambda| .$$

This is the assertion 10. The formula 9 follows now from 3.18.

Q.E.D.

6.4. Using the dictionary from 6.3, we may transform all results of §§ 3,4 to the assertions on the representations of symmetric groups. Let us mention three important classical results.

(1) The Littlewood-Richardson rule (Prop. 4.18'). It describes the restriction of irreducible representations of S_n to the subgroups $S_k \times S_l$ as well as the induction from $S_k \times S_l$ to S_n.

(2) The branching rule (Theorem 4.3). By the formula 6 from the table 6.3, this means that for any irreducible representation $\{\lambda\}$ of S_n the representation of S_{n-k} in the subspace of S_k'-

invariant vectors of $\{\lambda\}$ is multiplicity-free and equals

$$\sum\{\mu\} \qquad (\mu \perp \lambda \ , \ |\mu| = n-k).$$

In particular, when k=I we see that the restriction of $\{\lambda\}$ to S_{n-I} equals $\sum\{\mu\}$ where μ runs over all Young diagrams obtained by removing one point from the Young diagram λ.

(3) The Murnaghan-Nakayama character formula (Prop. 4.I7). By the formula 9 from the table 6.3 the inner product $\langle z_\nu, \{\lambda\}\rangle$ computed in 4.I7 (b) is the character value of the irreducible representation $\{\lambda\}$ at the class C_ν.

Another expression for character values is given by the Frobenius formula: if $\lambda, \nu \in \mathcal{P}_n$, $N \geqslant n$, and $L_I > L_2 > \dots > L_N \geqslant 0$ are the coordinates of λ described in 4.I6 then the character value of $\{\lambda\}$ at the class C_ν equals the coefficient of the monomial

$$\xi_1^{L_1}\xi_2^{L_2}\dots\xi_N^{L_N}$$

in the alternating polynomial

$$z_\nu(\xi_1,\dots,\xi_N)\cdot\Delta(\xi_1,\dots,\xi_N) \text{ (see 5.3, 5.I2).}$$

6.5. Now we compute dimensions of irreducible representations of the groups S_n.

<u>Proposition</u>. (a) Let $\lambda \in \mathcal{P}_n$ and $L_I > \dots > L_r$ be coordinates of λ described in 4.I6. Then

$$\dim\{\lambda\} = \frac{n!\prod_{i<j}(L_i - L_j)}{L_I!\,L_2!\,\dots\,L_r!}$$

(b) For any point $a = (i,j)$ belonging to the Young diagram λ the cardinality of the set

$$\{(i',j') \in \lambda \mid i' = i, j' \geqslant j \quad \text{or} \quad i' \geqslant i, j' = j\}$$

is called <u>hook length</u> of a and denoted by h(a). Then

$$\dim\{\lambda\} = \frac{n!}{\prod_{a\in\lambda} h(a)}$$

(this is the hook formula obtained in [11]).

<u>Proof</u>. (a) It follows from 6.3 (*) that the mapping, sending each $\pi \in \mathcal{A}(S_n)$ to $\dim \pi/|S_n| = \dim \pi/n!$, extends to the ring homomorphism

$$d : R(S) \longrightarrow \mathbb{Q} .$$

Applying d to the equality

$$\{\lambda\} = \det(x_{L_i-r+j}) \qquad \text{(see 4.16)},$$

we see that

$$\dim\{\lambda\} = n! \det\left(\frac{1}{(L_i-r+j)!}\right)$$

(we agree the convention that $1/(L_i-r+j)! = 0$ if $L_i-r+j < 0$). Multiplying the i-th row of this determinant by $L_i!$, we obtain that

$$\dim\{\lambda\} = \frac{n!}{L_1! \ldots L_r!} \det(P_{r-j}(L_i)),$$

where $P_k(L) = L(L-1)\ldots(L-k+1)$. Since P_k is a polynomial of degree k with the leading coefficient 1, the j-th column

$$(P_{r-j}(L_1),\ldots, P_{r-j}(L_r))$$

equals

$$(L_1^{r-j}, L_2^{r-j}, \ldots, L_r^{r-j}) + (\text{linear combination of previous columns}).$$

It follows that

$$\det (P_{r-j}(L_i)) = \det (L_i^{r-j}) = \prod_{i<j} (L_i - L_j)$$

(the well-known Vandermonde determinant), as desired.

(b) Under the notation of (a) it is easy to see that for any $i = 1,2,\ldots,r$ the number

$$L_i \,! \Big/ \prod_{j>i} (L_i - L_j)$$

is a product of hook lengths of all points of the i-th row of λ.

Q.E.D.

§ 7. Representations of wreath products

7.1. In this section we extend the results of § 6 to the representation theory of wreath products. Fix a finite group G. By the wreath product of G by S_n (the notation $S_n[G]$) we mean the semidirect product of S_n by G^n with respect to the action

$$\sigma : (g_1,\ldots,g_n) \longrightarrow (g_{\sigma^{-1}(1)}, \ldots, g_{\sigma^{-1}(n)}) \quad (\sigma \in S_n,\ g_i \in G).$$

In other words, elements of $S_n[G]$ are the expressions $\sigma \cdot (g_1,\ldots, g_n)$, where $\sigma \in S_n$, $g_i \in G$; the multiplication in $S_n[G]$ is defined by the commutation relation

$$(*) \qquad (g_1,\ldots,g_n)\cdot\sigma = \sigma\cdot(g_{\sigma(1)}, \ldots, g_{\sigma(n)}).$$

We consider S_n as a subgroup of $S_n[G]$, and G^n as a normal subgroup of $S_n[G]$. For any ordered partition $\alpha = (a_1,\ldots, a_r)$ of n we set

$$S_\alpha[G] = S_{a_1}[G] \times S_{a_2}[G] \times \ldots \times S_{a_r}[G]$$

and identify $S_\alpha[G]$ with the subgroup $S_\alpha \cdot G^n$ in $S_n[G]$ (see 6.3).

<u>Example</u>. If G consists of two elements, then $S_n[G]$ is the Weyl group of type C_n (or B_n); these groups are also called hyper octahedral. It is useful to keep this example in mind while reading this section.

7.2. Let us consider the graded group

$$R(S[G]) = \bigoplus_{n\geq 0} R(S_n[G]) \qquad \text{(see 6.I).}$$

It is a T-group with the set of irreducible elements

$$\coprod_{n\geq 0} \Omega(S_n[G]) \qquad \text{(see I.2, 6.I)}$$

We define the multiplication and comultiplication in $R(S[G])$ in a quite similar way as for symmetric groups. Namely, for $k+l=n$ we identify $R(S_k[G]) \otimes R(S_l[G])$ with $R(S_{(k,l)}[G])$ (see 6.I, 7.I) and define the multiplication

$$m : R(S_k[G]) \otimes R(S_l[G]) \longrightarrow R(S_n[G])$$

to be the induction from $S_{(k,l)}[G]$ to $S_n[G]$, and the comultiplication

$$m^* : R(S_n[G]) \longrightarrow \bigoplus_{k+l=n} (R(S_k[G]) \otimes R(S_l[G]))$$

as

$$\sum_{k+l=n} \operatorname{Res}^{S_n[G]}_{S_{(k,l)}[G]} .$$

<u>Proposition.</u> This setting makes $R(S[G])$ into a PSH-algebra. Its set of irreducible primitive elements is $\Omega(S_I[G]) = \Omega(G)$.

The proof is quite similar to that of Prop. 6.2 (the axiom

(H) will be verified in Appendix 3, A3.3).

Q.E.D.

7.3. By Theorem 2.2, $R(S[G])$ as a PSH-algebra decomposes into the tensor product of subalgebras $R(\rho)$, $\rho\in\Omega(G)$. By 3.I (g) each $R(\rho)$ as a PSH-algebra is isomorphic to the algebra R from §§ 3,4 (we will in this section represent R as $R(S)$ via the identification from 6.3). Let us describe $R(\rho)$ and define its identification with $R(S)$ more explicitly.

By definition (see 2.2), $R(\rho)_n$ for every $n\geqslant 0$ is spanned by irreducible constituents of ρ^n. We have

$$\rho^n = \operatorname{Ind}_{G^n}^{S_n[G]} (\rho\otimes\ldots\otimes\rho) \quad \text{(see 7.2)},$$

so by the Frobenius reciprocity $R(\rho)_n$ is spanned by irreducible representations of $S_n[G]$, whose restriction to the subgroup G^n contains $\rho\otimes\rho\otimes\ldots\otimes\rho$.

As in 6.3, an isomorphism of PSH-algebras $R(S)\xrightarrow{\sim} R(\rho)$ is uniquely determined by the image of x_2, the identity representation of S_2. We choose the isomorphism φ_ρ sending x_2 to the representation $\varphi_\rho(x_2)$ of $S_2[G] = S_2\cdot G^2$ such that

$$\varphi_\rho(x_2)\big|_{G^2} = \rho\otimes\rho \qquad \text{and} \qquad \varphi_\rho(x_2)\big|_{S_2} = \mathrm{id}.$$

Now we describe φ_ρ and the inverse isomorphism more explicitly from a functorial point of view. Let us define the additive functors

$$\varphi_\rho : \mathcal{A}(S_n) \longrightarrow \mathcal{A}(S_n[G]) \quad \text{and} \quad \Psi_\rho : \mathcal{A}(S_n[G]) \longrightarrow \mathcal{A}(S_n).$$

(see 6.I).

Let V be the space of ρ and $\otimes^n V$ be that of the representation $\otimes^n \rho = \rho \otimes \rho \otimes \ldots \otimes \rho$ of G^n. Define the representation τ of S_n in the space $\otimes^n V$ by

$$\tau(\sigma)(v_1 \otimes v_2 \otimes \ldots \otimes v_n) = v_{\sigma^{-1}(1)} \otimes v_{\sigma^{-1}(2)} \otimes \ldots \otimes v_{\sigma^{-1}(n)}$$

(1) Let $\pi \in \mathcal{A}(S_n)$ and W be the space of π. Define the representation $\bar{\pi} = \Phi_\rho(\pi) \in \mathcal{A}(S_n[G])$ in the space $W \otimes (\otimes^n V)$ by

$$\bar{\pi}(\sigma) = \pi(\sigma) \otimes \tau(\sigma) \qquad (\sigma \in S_n);$$

$$\bar{\pi}(g_1, \ldots, g_n) = \mathrm{id} \otimes (\rho(g_1) \otimes \ldots \otimes \rho(g_n)) \quad (g_i \in G).$$

($\bar{\pi}$ is well-defined according to 7.I ($*$)).

(2) Let $\bar{\pi} \in \mathcal{A}(S_n[G])$. Define the representation $\pi = \Psi_\rho(\bar{\pi}) \in \mathcal{A}(S_n)$ in the space $\mathrm{Hom}_{G^n}(\otimes^n \rho, \bar{\pi})$ by

$$\pi(\sigma)A = \bar{\pi}(\sigma) \circ A \circ \tau(\sigma^{-1})$$

(the inclusion $\pi(\sigma)A \in \mathrm{Hom}_{G^n}(\otimes^n \rho, \bar{\pi})$ follows from 7.I ($*$)).

Clearly, Φ_ρ and Ψ_ρ are additive functors. According to 6.I, we have the corresponding morphisms of graded T-groups

$$\Phi_\rho: R(S) \longrightarrow R(S[G]) \quad \text{and} \quad \Psi_\rho: R(S[G]) \longrightarrow R(S).$$

<u>Proposition.</u> (a) The operators Φ_ρ and Ψ_ρ are adjoint to each other with respect to the inner products $\langle\,,\,\rangle$.

(b) Each of Φ_ρ and Ψ_ρ is a PSH-algebra morphism i.e. a positive Hopf algebra morphism.

(c) The composition

$$\Psi_\rho \circ \Phi_\rho: R(S) \longrightarrow R(S)$$

is the identity operator, while

$$\varphi_\rho \circ \Psi_\rho : R(S[G]) \longrightarrow R(S[G])$$

is the orthogonal projection of $R(S[G])$ onto $R(\rho)$. In particular, φ_ρ and Ψ_ρ induce the mutually inverse isomorphisms of PSH-algebras $R(S)$ and $R(\rho)$.

Proof. (a) We will prove the more precise statement that the functor φ_ρ is adjoint of Ψ_ρ i.e. that for every $\pi \in \mathcal{A}(S_n)$, $\bar{\pi} \in \mathcal{A}(S_n[G])$ there exists a natural isomorphism

$$(*) \qquad \mathrm{Hom}_{S_n}(\pi, \Psi_\rho(\bar{\pi})) \xrightarrow{\sim} \mathrm{Hom}_{S_n[G]}(\varphi_\rho(\pi), \bar{\pi}).$$

Let W be the space of π, and $A \in \mathrm{Hom}_{S_n}(\pi, \Psi_\rho(\bar{\pi}))$. By definition, A is an operator from W to $\mathrm{Hom}_{G^n}(\otimes^n \rho, \bar{\pi})$. Define the operator $\bar{A} : W \otimes (\otimes^n V) \longrightarrow \bar{\pi}$ by

$$\bar{A}(w \otimes v_I \otimes \ldots \otimes v_n) = A(w)(v_I \otimes \ldots \otimes v_n).$$

One can easily verify that $\bar{A} \in \mathrm{Hom}_{S_n[G]}(\varphi_\rho(\pi), \bar{\pi})$ and the correspondence $A \longmapsto \bar{A}$ establishes $(*)$.

(b) Clearly, φ_ρ and Ψ_ρ are positive; definitions imply at once that they are coalgebra morphisms. It is rather tedious to verify directly that φ_ρ and Ψ_ρ are ring homomorphisms, but this follows at once from (a) and the axiom (S) from I.4: since φ_ρ (Ψ_ρ) is a coalgebra morphism, its adjoint Ψ_ρ (φ_ρ) is a ring homomorphism.

(c) By (a), for every $\pi \in \mathcal{A}(S_n)$ we have the natural morphism

$$A : \pi \longrightarrow \Psi_\rho \circ \varphi_\rho(\pi)$$

(apply $(*)$ for $\bar{\pi} = \varphi_\rho(\pi)$ and choose A corresponding to the identity morphism at the right-hand side of $(*)$). Similarly, for every $\bar{\pi} \in \mathcal{A}(S_n[G])$ we have the natural morphism

$$\bar{A} : \mathcal{P}_\rho \circ \Psi_\rho (\bar{\pi}) \longrightarrow \bar{\pi}$$

It suffices to prove that A is an isomorphism while $\bar{A}$ maps $\mathcal{P}_\rho \circ \bar{\Psi}_\rho (\bar{\pi})$ isomorphically onto the sum of all constituents of $\bar{\pi}$, belonging to $R(\rho)$. The straightforward proof is left to the reader.

Q.E.D.

7.4. Combining 7.3 and 4.19, we obtain the classification of irreducible representations of the groups $S_n[G]$ in terms of that of G. Namely, irreducible representations of $S_n[G]$ are parametrized by the set $S_n(\Omega(G); \mathcal{P})$ (see 4.19) via the formula

$$\varphi \longmapsto \{\varphi\} = \prod_{\rho \in \Omega(G)} \mathcal{P}_\rho(\{\varphi(\rho)\})$$

7.5. By the isomorphisms $\mathcal{P}_\rho$ we transfer the operator $t:R(S) \longrightarrow R(S)$ (see 3.11) to all $R(\rho)$ and then extend it to the involutive automorphism

$$t:R(S[G]) \longrightarrow R(S[G]).$$

Clearly, t acts on irreducible representations by

$$t(\{\varphi\}) = \{\varphi^t\} \quad , \text{ where } \quad \varphi^t(\rho) = \varphi(\rho^t)$$

(see 7.4, 4.2).

<u>Proposition</u>. If $\pi \in \mathcal{A}(S_n[G])$ then $t(\pi) = \pi \otimes \varepsilon_S$, the (inner) tensor product of π and the character ε_S of $S_n[G]$ which equals ε on S_n and 1 on G^n.

This follows at once from the description of t in R(S) (the formula 5 from the table 6.3), the explicit construction of $\mathcal{P}_\rho$ in 7.3, and the obvious fact that the operator $\pi \longmapsto \pi \otimes \varepsilon_S$ extends to an algebra morphism $R(S[G]) \longrightarrow R(S[G])$.

Q.E.D.

7.6. Proposition 4.18 allows us to compute explicitly the action of multiplication and comultiplication in $R(S[G])$ on irreducible representations. In other words, we know the induction from $S_\alpha[G]$ to $S_n[G]$ and the restriction from $S_n[G]$ to $S_\alpha[G]$. For example, we obtain the following.

Proposition. (Cf. 6.4 (2)). The restriction of an irreducible representation $\{\varphi\}$ of $S_n[G]$ (see 7.3) to the subgroup $S_{(n-1,1)}[G] = S_{n-1}[G] \times G$ is multiplicity-free and equals

$$\sum_{\rho\in\Omega(G)} \sum_{\varphi_\rho} \{\varphi_\rho\} \otimes \rho ,$$

where the inner sum is over $\varphi_\rho \in S_{n-1}(\Omega(G); \mathcal{P})$ such that $\varphi_\rho(\rho') = \varphi(\rho')$ for $\rho' \neq \rho$, and $\varphi_\rho(\rho)$ is obtained by removing one point from the Young diagram $\varphi(\rho)$.

7.7. Now we consider the characters of the groups $S_n[G]$. Our goal is to construct the character table of $S_n[G]$ in terms of that of G and of the groups S_k (see 6.4 (3)). First we describe conjugacy classes of $S_n[G]$.

As in § 6 (see 6.1) we identify $R(S[G])$ with the lattice in $C(S[G]) = \bigoplus_{n\geq 0} C(S_n[G])$, and consider $C(S[G])$ as a Hopf algebra over $\mathbb{C}$. To each conjugacy class C of $S_n[G]$ we assign the function

$$\zeta_C = |S_n[G]|/|C| \cdot \chi_C \quad \in C(S_n[G]),$$

where χ_C is the characteristic function of C. We say that C is <u>primitive</u> if ζ_C is a primitive element of the Hopf algebra $C(S[G])$; clearly, this means that C intersects trivially any subgroup $S_{(k,l)}[G]$ for $k,l>0$, $k+l=n$. Denote by $K(G)$ the set of conjugacy classes of G (it is graded by $\deg \equiv 1$; see 4.19).

<u>Proposition.</u> (a) For each $n \geqslant 1$ primitive conjugacy classes in $S_n[G]$ are parametrized by the set $K(G)$: the primitive class $C(n;K)$ in $S_n[G]$, corresponding to a class $K \in K(G)$, contains all elements of the form $\sigma_n \cdot (g_1, g_2, \ldots, g_n)$, where

$$\sigma_n = (n \longrightarrow n-1 \longrightarrow n-2 \longrightarrow \ldots \longrightarrow 1 \longrightarrow n) \in S_n$$

is an n-cycle, and

$$g_1 \cdot g_2 \cdot \ldots \cdot g_n \in K .$$

We shall write $\zeta_{n,K}$ for $\zeta_{C(n;K)}$.

(b) Conjugacy classes in $S_n[G]$ are parametrized by the set $S_n(K(G); \mathcal{P})$ (see 4.19). The class $C(\psi)$ corresponding to a function $\psi : K(G) \longrightarrow \mathcal{P}$, is defined by

$$\zeta_{C(\psi)} = \prod_{K \in K(G)} \prod_{k \in \mathbb{N}} \zeta_{l_k(K),K} ,$$

where $\psi(K) = (l_1(K), l_2(K), \ldots)$. In other words, $C(\psi)$ contains the conjugacy class

$$\prod_K \prod_k C(l_k(K); K)$$

of the subgroup

$$\prod_K \prod_k S_{l_k(K)}[G] \subset S_n[G]$$

(see 6.3 (*)).

<u>Proof.</u> (a) It is easy to see that every primitive conjugacy class in $S_n[G]$ consists of elements of the form $\sigma \cdot (g_1, \ldots, g_n)$, where σ is an n-cycle in S_n. Evidently, every such element is conjugate to one of the form $\sigma_n \cdot (g_1, \ldots, g_n)$ under transformation by an element of S_n. It remains to find the orbits of the action of $Z_{S_n}(\sigma_n) \cdot G^n$ by inner automorphisms on the set $\sigma_n \cdot G^n$ (here

$Z_{S_n}(\sigma_n)$ is the centralizer of σ in S_n; it is easy to see that it is the cyclic subgroup $\langle\sigma_n\rangle$ generated by σ_n).

Define the mapping $\prod : \sigma_n \cdot G^n \longrightarrow G$ by

$$\prod(\sigma_n \cdot (g_I, g_2, \ldots, g_n)) = g_I g_2 \ldots g_n$$

To prove our statement it suffices to verify that two elements g and h of $\sigma_n \cdot G^n$ are conjugate under the action of $\langle\sigma_n\rangle \cdot G^n$ if and only if $\prod(g)$ and $\prod(h)$ are conjugate in G.

By 7.I (*) we have

$$(*) \qquad \sigma_n \cdot (\sigma_n(g_I, \ldots, g_n)) \cdot \sigma_n^{-I} = \sigma_n \cdot (g_2, g_3, \ldots, g_n, g_I);$$

$$(**) \qquad (h_I, \ldots, h_n) \cdot (\sigma_n \cdot (g_I, \ldots g_n)) \cdot (h_I, \ldots, h_n)^{-I} =$$

$$= \sigma_n \cdot (h_n g_I h_I^{-I}, h_I g_2 h_2^{-I}, h_2 g_3 h_3^{-I}, \ldots, h_{n-I} g_n h_n^{-I}).$$

It follows that for $g = \sigma_n \cdot (g_I, \ldots, g_n) \in \sigma_n \cdot G^n$ we have

$$\prod(\sigma_n g \sigma_n^{-I}) = g_I^{-I} \cdot \prod(g) \cdot g_I ,$$

$$\prod((h_I, \ldots, h_n)\, g\, (h_I, \ldots, h_n)^{-I}) = h_n \cdot \prod(g) \cdot h_n^{-I}$$

This proves the part ծnly if" of the above statement. The part "if" also follows readily from (**).

(b) The shortest way is to apply the structural theory of Hopf algebras over fields of characteristic zero (see 2.I). By (a), the elements $\zeta_{l,K}$ ($l \geq I$, $K \in K(G)$) form a basis of the space of primitive elements in $C(S[G])$. According to 2.I (I), (2), $C(S[G])$ is a polynomial ring $\mathbb{C}[\{\zeta_{l,K}\}]$ (see also Appendix I, AI.I-AI.2). This means that the elements $\zeta_{C(\psi)}$ ($\psi \in S(K(G); \mathcal{P})$) form a basis in $C(S[G])$.

It follows that the conjugacy classes $C(\psi)$ are pairwise distinct and exhaust conjugacy classes of all groups $S_n[G]$, as desired.

Let us give another proof. The elements $\zeta_{1,K}$ are primitive and mutually orthogonal, so by 2.4 the elements $\zeta_{C(\psi)}$ are mutually orthogonal. Hence all classes $C(\psi)$ are distinct. But by 7.4 the number of such classes in $S_n[G]$ equals the number of irreducible representations of $S_n[G]$.

Q.E.D.

7.8. By definition, the character value of an irreducible representation $\{\varphi\}$ of $S_n[G]$ at a conjugacy class $C(\psi)$ (see 7.4, 7.7) equals the inner product

$$<\{\varphi\}, \zeta_{C(\psi)}>.$$

For any $\varphi \in S(\Omega(G);\mathcal{P})$ set

$$z_\varphi = \prod_{\rho \in \Omega(G)} \mathcal{P}_\rho(z_{\varphi(\rho)}) \in R(S[G]) \otimes \mathbf{Q} \subset C(S[G]).$$

(see 3.16, 7.3). According to 7.3 we have

$$<\{\varphi\}, z_{\varphi'}> = \prod_{\rho \in \Omega(G)} <\{\varphi(\rho), z_{\varphi'(\rho)}> ;$$

the right-hand side is known from 4.17. We see that to find the character table of each $S_n[G]$ it suffices to compute the transition matrix between the bases $\{\zeta_{C(\psi)}\}$ and $\{z_\varphi\}$ in $C(S[G])$.

Fix $n \geq 1$. It is easy to see that each of the sets

$$\{\mathcal{P}_\rho(z_n) \mid \rho \in \Omega(G)\} \quad \text{and} \quad \{\zeta_{n,K} \mid K \in K(G)\}$$

is a basis of the space of primitive elements in $C(S_n[G])$ (see 3.15, 7.3, and 7.7). We shall compute the transition matrix between these bases (evidently, this enables us to compute the transition matrix between the bases $\{z_\varphi\}$ and $\{\zeta_{C(\psi)}\}$. It turns out that this matrix does not depend on n and equals simply the character table of G.

Proposition. For every $\rho \in \Omega(G)$ and $K \in K(G)$ denote by $\rho(K)$ the character value of the representation ρ at the conjugacy class K. Then for $n \geqslant I$ we have

$$\zeta_{n,K} = \sum_{\rho \in \Omega(G)} \overline{\rho(K)} \cdot \Phi_\rho(z_n);$$
$$\Phi_\rho(z_n) = \sum_{K \in K(G)} \rho(K) \cdot |K|/|G| \cdot \zeta_{n,K}$$

(here "——" stands for complex conjugation).

Proof. We shall show that for every $\pi \in C(S_n)$,

(∗) $\quad < \Phi_\rho(\pi), \zeta_{n,K} > = <\pi, z_n> \cdot \rho(K)$.

It suffices to consider the case when π is an ordinary representation of S_n. In this case the left-hand side of (∗) is the character value of $\Phi_\rho(\pi)$ at $C(n;K)$, i.e. the trace of the operator

$$X = \Phi_\rho(\pi) \; (\sigma_n \cdot (k,e,e,\ldots,e)),$$

where $k \in K$ (see 7.7). Under the notation of 7.3 (I), X acts on the space $W \otimes (\otimes^n V)$ by

$$X \, (w \otimes v_I \otimes \ldots \otimes v_n) = \pi(\sigma_n) \, w \otimes v_2 \otimes v_3 \otimes \ldots \otimes v_n \otimes \rho(k) \, v_I \; .$$

The trace of this transformation can be computed directly, and we obtain:

$$\operatorname{tr} X = \operatorname{tr} \pi(\sigma_n) \cdot \operatorname{tr} \rho(k) = <\pi, z_n> \cdot \rho(K)$$

(see 6.3), as desired.

Evidently, the elements $\Phi_\rho(z_n)$ for $\rho \in \Omega(G)$ are mutually orthogonal (since they belong to the distinct $R(\rho)$). Therefore

$$\zeta_{n,K} = \sum_{\rho \in \Omega(G)} \frac{\langle \zeta_{n,K}, \Phi_\rho(z_n)\rangle}{\langle \Phi_\rho(z_n), \Phi_\rho(z_n)\rangle} \cdot \Phi_\rho(z_n) .$$

Substituting $\pi = z_n$ into (*) and using that Φ_ρ is an isometry, we obtain the first formula to be proved. The second formula follows from the first one and the orthogonality relations on G.

Q.E.D.

7.9. The theory is considerably simplified in the case when G is abelian. The construction of the functors Φ_ρ and Ψ_ρ is simplified by the fact that all irreducible representations of G are one-dimensional (see 7.3). The classification of irreducible representations of $S_n[G]$ (see 7.4) in this case can be derived directly from the general representation theory of groups with an abelian normal subgroup (see [9], 9.2). Moreover, when G is abelian, some of results become more beautiful. For example, we have the following

<u>Branching Rule</u> (cf. 6.4(2)). If G is abelian then the restriction of any irreducible representation $\{\varphi\}$ of $S_n[G]$ to the subgroup $S_{n-1}[G]$ is multiplicity-free and equals

$$\sum_{\rho \in \Omega(G)} \sum_{\varphi_\rho} \{\varphi_\rho\}$$

(the φ_ρ are as in Prop. 7.6).

<u>Proof</u>. By 7.6, this restriction always equals

$$\sum_{\rho} \sum_{\varphi_\rho} \dim \rho \cdot \{\varphi_\rho\}$$

(even if G is not supposed to be abelian). If G is abelian then $\dim\rho = 1$ for all $\rho \in \Omega(G)$.

Q.E.D.

7.10. Now we compute the restriction of representations of $S_n[G]$ to S_n and the induction from S_n to $S_n[G]$. We define the graded group morphisms

$$I: R(S) \longrightarrow R(S[G]) \quad \text{and} \quad R: R(S[G]) \longrightarrow R(S)$$

to be

$$I = \mathrm{Ind}_{S_n}^{S_n[G]} \quad \text{on each } R(S_n) \text{ and}$$

$$R = \mathrm{Res}_{S_n}^{S_n[G]} \quad \text{on each } R(S_n[G]).$$

<u>Proposition</u>. (a) The operators I and R are adjoint to each other and are both PSH-algebra morphisms (i.e. positive Hopf algebra morphisms).

(b) Suppose G to be abelian. Then for any $\rho \in \Omega(G)$ we have

$$R \circ \Phi_\rho = \Psi_\rho \circ I = \mathrm{id} : R(S) \longrightarrow R(S) \quad \text{(see 7.3)}.$$

In other words, if we identify each $R(\rho)$ with $R(S)$ as in 7.3 and so identify $R(S[G]) = \bigotimes_\rho R(\rho)$ with $\bigotimes_\rho R(S)$ then the operator

$$I : R(S) \longrightarrow \bigotimes_\rho R(S)$$

is induced by comultiplication in $R(S)$ (more precisely, it is the operator P_I from 5.2) while

$$R : \bigotimes_\rho R(S) \longrightarrow R(S)$$

is induced by multiplication in $R(S)$. It is amusing that multiplication is now realized as restriction while comultiplication as induction.

Proof. (a) The adjointness of I and R follows at once from the Frobenius reciprocity. Evidently I and R are positive and R is a coalgebra morphism. The fact that R is an algebra morphism, can be derived readily from the Mackey theorem on the restriction of induced representations ([9], 7.4; see also Appendix 3, A3.4 below). Since I and R are adjoint to each other, the axiom (S) from I.4 implies that I as well as R is a Hopf algebra morphism.

(b) Since ρ is one-dimensional, the equality $R \circ \Phi_\rho = \mathrm{id}$ follows at once from the construction of Φ_ρ (see 7.3). The equality $\Psi_\rho \circ I = \mathrm{id}$ is obtained by passing to adjoint operators (see 7.3 (a)).

Q.E.D.

7.II. In conclusion, we compute dimensions of irreducible representations of $S_n[G]$.

Proposition. The dimension of an irreducible representation $\{\varphi\}$ of $S_n[G]$ equals

$$\frac{n!}{\prod_{\rho \in \Omega(G)} \prod_{a \in \varphi(\rho)} h(a)} \cdot \prod_{\rho \in \Omega(G)} (\dim \rho)^{|\varphi(\rho)|} \qquad \text{(see 6.5 (b))}$$

(note that when G is abelian, the second factor is 1).

This follows at once from 7.3, 7.4, and 6.5 (b).

Q.E.D.

Chapter III. Representations of general linear and affine groups over finite fields

§ 8. Functors $i_{U,\theta}$ and $r_{U,\theta}$

8.1. In this section we introduce the functors generalizing induction and restriction of representations, and outline their main properties. Let G be a finite group, and M and U its subgroups such that M normalizes U and $M \cap U = \{e\}$. Let $\theta : U \longrightarrow \mathbb{C}^*$ be a character of U normalized by M, i.e. such that

$$\theta(mum^{-1}) = \theta(u) \quad \text{for} \quad m \in M,\ u \in U.$$

In this setting we define the functors

$$i_{U,\theta} : \mathcal{A}(M) \longrightarrow \mathcal{A}(G) \qquad \text{("}\theta\text{-induction")}$$

and

$$r_{U,\theta} : \mathcal{A}(G) \longrightarrow \mathcal{A}(M) \qquad \text{("}\theta\text{-restriction")}$$

(see 6.1).

(1) Let V be the space of $\rho \in \mathcal{A}(M)$. We extend ρ to the representation $\bar{\rho}$ of $P = MU$ in the same space V, setting $\bar{\rho}(u) = \theta(u) \cdot 1_V$ for $u \in U$. Put

$$i_{U,\theta}(\rho) = \operatorname{Ind}_P^G(\bar{\rho}).$$

In other words, $i_{U,\theta}(\rho)$ is the representation of G by right translations in the space of functions $f : G \longrightarrow V$ such that

$$f(mug) = \theta(u) \cdot \rho(m) f(g) \quad \text{for} \quad m \in M,\ u \in U,\ g \in G.$$

(2) Let E be the space of $\pi \in \mathcal{A}(G)$. We set

$$E^{U,\theta} = \{\xi \in E \mid \pi(u)\xi = \theta(u)\cdot\xi \quad \text{for all} \quad u \in U\}$$

Clearly, the subspace $E^{U,\theta} \subset E$ is M-invariant. By definition, $r_{U,\theta}(\pi)$ is the representation of M in $E^{U,\theta}$.

When $U = \{e\}$, the functors $i_{U,\theta}$ and $r_{U,\theta}$ become usual induction and restriction. Their main properties remain valid in the general situation.

Proposition. (a) The functors $i_{U,\theta}$ and $r_{U,\theta}$ are additive.

(b) $r_{U,\theta}$ is adjoint of $i_{U,\theta}$, i.e. for any $\rho \in \mathcal{A}(M)$, $\pi \in \mathcal{A}(G)$ we have a natural isomorphism

$$\text{Hom}\,(r_{U,\theta}(\pi), \rho) = \text{Hom}\,(\pi, i_{U,\theta}(\rho)).$$

(c) Let N and V be subgroups of M and θ' be a character of V such that the functors

$$i_{V,\theta'}: \mathcal{A}(N) \longrightarrow \mathcal{A}(M) \quad \text{and} \quad r_{V,\theta'}: \mathcal{A}(M) \longrightarrow \mathcal{A}(N)$$

make sense. Define the character θ^o of $U^o = UV$ by

$$\theta^o(uv) = \theta(u)\cdot\theta'(v), \quad u \in U, \quad v \in V .$$

Then

$$i_{U,\theta} \circ i_{V,\theta'} = i_{U^o,\theta^o}\,, \quad r_{V,\theta'} \circ r_{U,\theta} = r_{U^o,\theta^o}$$

(d) If H is a finite group, and $\tau \in \mathcal{A}(H)$ then each of $i_{U,\theta}$ and $r_{U,\theta}$ commutes with the functors $* \longmapsto * \otimes \tau$ and $* \longmapsto \tau \otimes *$.

Proof. Part (b) follows readily from the Frobenius reciprocity while other assertions are immediate consequences of the definitions.

Q.E.D.

8.2. Since the functor $i_{U,\theta} : \mathcal{A}(M) \longrightarrow \mathcal{A}(G)$ is additive, it induces the positive morphism

$$i_{U,\theta} : R(M) \longrightarrow R(G)$$

and the $\mathbb{C}$ - linear operator

$$i_{U,\theta} : C(M) \longrightarrow C(G)$$

(the similar holds for $r_{U,\theta}$). All assertions of Prop. 8.I can be obviously reformulated in terms of these operators.

Let us give explicit formulas for the action of $i_{U,\theta}$ and $r_{U,\theta}$ on class functions.

<u>Proposition</u>. Under the notation of 8.I we have for $\chi \in C(M)$, $g \in G$

$$i_{U,\theta}(\chi)\ (g) = \frac{1}{|M|\cdot|U|} \sum \chi\,(m)\cdot\theta(u)$$

(the sum is over the set $\{ (g_1, m, u) \in G \times M \times U \mid g = g_1\, m u g_1^{-1} \}$) and for $\varphi \in C(G)$, $m \in M$

$$r_{U,\theta}\ (\varphi)\ (m) = \frac{1}{|U|} \sum_{u \in U} \theta^{-1}\ (u)\cdot\varphi\,(mu)$$

The proof is left to the reader.

8.3. Let M,U,N, ans V be subgroups of a group G, θ be a character of U and ψ be a character of V such that the functors

$$i_{U,\theta} : \mathcal{A}(M) \longrightarrow \mathcal{A}(G) \quad \text{and} \quad r_{V,\psi} : \mathcal{A}(G) \longrightarrow \mathcal{A}(N)$$

make sense. Under some extra assumptions one can compute the composition

$$r_{V,\psi} \circ i_{U,\theta} : \mathcal{A}(M) \longrightarrow \mathcal{A}(N) .$$

Since the formulation is rather cumbersome, it will be given in Appendix 3 (when $U = V = \{e\}$ one obtains the Mackey theorem; see [9], 7.4). The computation of such a composition in various situations will be one of our main tools. In the body of the text the results of this kind will be only formulated while the proofs will be given in Appendix 3.

§ 9. The classification of irreducible representations of $GL(n, \mathbb{F}_q)$

9.I. Fix a finite field $\mathbb{F}_q$. We shall write G_n for $GL(n, \mathbb{F}_q)$ and $R(q)$ for $\bigoplus_{n \geq 0} R(G_n)$ (see 6.I); thus, $R(q)$ is a graded T-group (we agree the convention that $G_0 = \{e\}$ so $R(q)_0 = R(G_0) = \mathbb{Z}$). We shall make $R(q)$ into a Hopf algebra.

As in I.I, we must for all k,l,n with $k+l=n$ define multiplication

$$m : R(G_k \times G_l) \longrightarrow R(G_n)$$

and the component of comultiplication

$$m^*_{k,l} : R(G_n) \longrightarrow R(G_k \times G_l) .$$

Let us embed $M = G_k \times G_l$ into G_n as the subgroup of cellular diagonal matrices and consider the subgroup $U = U_{k,l} \subset G_n$ consisting of unipotent upper triangular matrices $u = (u_{ij})$ such that u_{ij} can be non-zero only if $i = j$ or $i \leq k < j$. Clearly, M normalizes U and $M \cap U = \{e\}$, so there are defined the functors

$$i_{U,1} : \mathcal{A}(M) \longrightarrow \mathcal{A}(G_n) \quad \text{and} \quad r_{U,1} : \mathcal{A}(G_n) \longrightarrow \mathcal{A}(M)$$

(see 8.I). We put

$$m = i_{U,1} ,\ m^*_{k,l} = r_{U,1} \quad \text{and}$$

$$m^{*}/_{R(G_n)} = \sum_{k+l=n} m^{*}_{k,l}$$

<u>Proposition</u>. This setting makes $R(q)$ into a PSH-algebra.

<u>Proof</u>. The proof is just the same as that of Prop. 6.2 with the unique exception that the axiom (H) is now derived not from the Mackey theorem but from the theorem on composition of the functors $r_{V,\psi}$ and $i_{U,\theta}$ (see 8.3); this is done in Appendix 3, A3.5 below.

Q.E.D.

Note that the commutativity of multiplication (and comultiplition) in $R(q)$, which in our approach is obtained automatically, is a non-trivial property of representations of G_n; all other proofs known by the author are based on the antiautomorphism of transposition of matrices in G_n (see [4],[7],[15]).

9.2. Irreducible primitive elements of $R(q)$ are called <u>cuspidal representations</u> *). By definition, a representation $\rho \in \Omega(G_n)$ is cuspidal if and only if it contains no non-zero vectors invariant under some non-trivial subgroup $U_{k,l}$. In terms of characters this means that the character of ρ is a cusp form on G_n (see [24]).

9.3. Denote by $\mathcal{C}$ the set of irreducible primitive elements of $R(q)$ (i.e. the set of equivalence classes of cuspidal representations of the groups G_n). By Theorem 2.2, $R(q)$ decomposes into the tensor product $\bigotimes_{\rho\in\mathcal{C}} R(\rho)$. After D.K.Faddeev [7] an irreducible element of $R(\rho)$ (i.e. an irreducible constituent of some

---

*) Some authors call them representations of discrete series ([15]), analitic ([14]) or prime ([7]); it is natural for us to call them primitive but we prefer not to enlarge this list.

power of ρ)will be called a primary representation corresponding to ρ . The equality

$$R(q) = \bigotimes_{\rho\in\mathcal{C}} R(\rho)$$

means that any irreducible representation of G_n decomposes uniquely up to a permutation into the product of primary ones corresponding to distinct cuspidal representations.

9.4. According to 4.19, if we choose for any $\rho\in\mathcal{C}$ a PSH-algebra isomorphism

$$\varphi_\rho : R \xrightarrow{\sim} R(\rho)$$

then irreducible representations of any group G_n would be parametrized by the set $S_n(\mathcal{C};\mathcal{P})$(here $\mathcal{C}$ is naturally graded by

$$\deg\rho = m \qquad \text{for} \qquad \rho\in\Omega(G_m) \;).$$

To construct the φ_ρ we shall use Prop. 3.6; so we want to construct for any $\rho\in\mathcal{C}$ a linear form $\delta : R(\rho)\longrightarrow\mathbb{Z}$ satisfying hypothesis of Prop. 3.6. Fix a non-trivial additive character

$$\psi : \mathbb{F}_q \longrightarrow \mathbb{C}^* .$$

We shall denote by the same letter ψ the character of the group $U = U_n\subset G_n$ of unipotent upper triangular matrices, sending $u = (u_{ij})$ to

$$\psi\Big(\sum_{1\le i<n} u_{i,i+1}\Big),$$

as well as its restriction to various subgroups of U. For every n define the linear form

$$\delta : R(G_n)\longrightarrow R(\{e\}) = \mathbb{Z}$$

by $\delta = r_{U,\psi}$ (see 8.1);

thus, we have constructed the form $\delta : R(q) \longrightarrow \mathbb{Z}$.

Proposition. δ is positive, multiplicative, and $\delta(\rho) = 1$
for all $\rho \in \mathcal{C}$.

Proof. Obviously, δ is positive. To verify that δ is multiplicative one must compute certain composition of the form
$r_{V,\psi} \circ i_{U,\theta}$; this will be done as usual in Appendix 3 (see A3.6).
The most deep assertion is that $\delta(\rho) = I$. It consists of two
parts:

(I) $\delta(\omega) \leq I$ for any $\omega \in \Omega(G_n)$;

(2) $\delta(\rho) > 0$ for any cuspidal ρ .

Part (I) is due to I.M.Gelfand and M.I.Graev ([I3]), while (2) is
due to S.I.Gelfand ([I4]).

Q.E.D.

9.5. By 3.6 and 9.4 if $\Phi_\rho : R \longrightarrow R(\rho)$ is a PSH-algebra isomorphism then the form $\delta \circ \Phi_\rho : R \longrightarrow \mathbb{Z}$ equals either
δ_x or δ_y . We choose for any $\rho \in \mathcal{C}$ the isomorphism Φ_ρ
such that

$$\delta \circ \Phi_\rho = \delta_y .$$

Let us give another description of Φ_ρ . Following I.M.Gelfand
and M.I.Graev [I3] , we say that a representation ω of G_n is
non-degenerate if $\delta(\omega) \neq 0$, and degenerate if $\delta(\omega) = 0$. By
9.4, for $\rho \in \mathcal{C}$ and $n \geq I$ the representation ρ^n has the unique non-degenerate irreducible constituent ω ; the isomorphism
is uniquely determined by the condition

$$\Phi_\rho(y_n) = \omega .$$

In § II (II.2-II.6) we shall give the independent proof of

the equality $\delta(\rho) = 1$ (see 9.4) i.e. of the theorems by I.M.Gelfand and M.I.Graev, and by S.I.Gelfand. To avoid a vicious circle we shall not use this equality (and so the explicit choise of φ_ρ) till II.6. Now we discuss the important special case when the equality $\delta(\rho) = 1$ becomes obvious; this example will play the crucial role in § 10.

9.6. By definition any irreducible representation of the group G_1 is cuspidal (since $G_1 = \mathbb{F}_q^*$, these representations are simply multiplicative characters of $\mathbb{F}_q$). In particular, let us consider the identity representation ι of G_1 and the subalgebra $R(\iota) \subset R(q)$. It is easy to see that irreducible elements of $R(\iota)$ are irreducible representations of the groups G_n containing a non-zero vector invariant under the upper triangular subgroup B_n. In particular, the identity representation ι_n of G_n lies in $R(\iota)$.

The equality $\delta(\iota) = 1$ follows at once from definitions (see 9.4, 8.1). So by 9.5 we obtain the well-defined isomorphism

$$\varphi_\iota : R \xrightarrow{\sim} R(\iota)$$

(note that the definition of φ_ι doesn't need Proposition 9.4 and so the theorems by I.M.Gelfand and M.I.Graev, and S.I.Gelfand). The next proposition justifies this definition, i.e. explains why it is natural to identify δ with δ_y, not with δ_x.

<u>Proposition</u>. We have

$$\varphi_\iota(x_n) = \iota_n \quad .$$

In other words, identifying R with $R(S)$ as in 6.3, we see that φ_ι sends the identity representation of S_n to that of G_n.

<u>Proof</u>. By definition, $\delta(\iota_n) = 0$ for $n \geq 2$. Evidently,

$$m^*(\iota_n) = \sum_{k=0}^{n} \iota_k \otimes \iota_{n-k} \; ,$$

By I.9 (a), (b):

$$y_2^*(\varphi_\iota^{-1}(\iota_n)) = 0 \quad \text{for all } n,$$

hence by 3.I (b), $\iota_n = \varphi_\iota(x_n)$.

Q.E.D.

9.7. Remark. The isomorphism between $R(\iota)$ and $R(S)$ is due to R.Steinberg [25] ; the non-degenerate irreducible constituent of the representation

$$\iota^n = \mathrm{Ind}_{B_n}^{G_n} 1 ,$$

i.e.the element $\varphi_\iota(y_n)$ is the well-known Steinberg representation.

§ IO. The P.Hall algebra

IO.I. Denote by $\mathcal{H}_n$ the subspace in $C(G_n)$ (see 6.I) consisting of functions with support in the set of unipotent elements of G_n. Put

$$\mathcal{H} = \bigoplus_{n\geq 0} \mathcal{H}_n ,$$

so $\mathcal{H}$ is a graded subspace of

$$\bigoplus_{n\geq 0} C(G_n) = R(q) \otimes \mathbb{C} \qquad \text{(see 6.I, 9.I)} .$$

Let

$$p : R(q) \otimes \mathbb{C} \longrightarrow \mathcal{H}$$

be the operator of orthogonal projection onto $\mathcal{H}$; in other words, if $f \in C(G_n)$ then $p(f)$ is the restriction of f to the set of unipotent elements of G_n .

By the theory of the Jordan normal form, unipotent conjugacy classes of the groups G_n are parametrized by partitions: to a partition $\lambda = (l_1,..,l_r) \in \mathcal{P}$ there corresponds the class K_λ consisting of elements with Jordan cells of orders $l_1, l_2, \dots, l_r$. Denote by $\chi_\lambda \in \mathcal{H}$ the characteristic function of K_λ, so $\{\chi_\lambda \mid \lambda \in \mathcal{P}\}$ is a basis of $\mathcal{H}$. For $\lambda \in \mathcal{P}_n$ we put

$$|Z(K_\lambda)| = |G_n|/|K_\lambda| .$$

In the next section we shall compute explicitly the map p, and so the character values of G_n at unipotent elements. Now we investigate $\mathcal{H}$ in detail.

Proposition. (a) $\mathcal{H}$ is a Hopf subalgebra in $R(q) \otimes \mathbb{C}$, and $p : R(q) \otimes \mathbb{C} \longrightarrow \mathcal{H}$ is a Hopf algebra morphism. We call $\mathcal{H}$ the Hall algebra.

(b) Let $g^\lambda_{\mu\nu}$ and $g^{\mu\nu}_\lambda$ be structural constants of $\mathcal{H}$, given by

$$\chi_\mu \cdot \chi_\nu = \sum_{\lambda \in \mathcal{P}} g^\lambda_{\mu\nu} \cdot \chi_\lambda \quad ;$$

$$m^*(\chi_\lambda) = \sum_{\mu,\nu} g^{\mu\nu}_\lambda \cdot \chi_\mu \otimes \chi_\nu .$$

Then

$$g^{\mu\nu}_\lambda = \frac{|Z(K_\mu)| \cdot |Z(K_\nu)|}{|Z(K_\lambda)|} \cdot g^\lambda_{\mu\nu} ,$$

while $g^\lambda_{\mu\nu}$ can be computed as follows. Let $|\lambda| = n$ and $u \in K_\lambda$; then $g^\lambda_{\mu\nu}$ equals the number of subspaces $V \subset \mathbb{F}_q^n$ invariant under u and such that the action of u on V belongs to the class K_μ, while the action of u on $\mathbb{F}_q^n/V$ belongs to K_ν.

Proof. (a) The definition of multiplication and comultiplication on $R(q) \otimes \mathbb{C}$ in terms of characters (see 9.1, 8.2) implies at once that $\mathcal{H}$ is a Hopf subalgebra, while its orthogonal complement $\mathcal{H}^\perp$ is an ideal in $R(q) \otimes \mathbb{C}$ such that

$$m^*(\mathcal{H}^\perp) \subset (\mathcal{H}^\perp \otimes R(q)_{\mathbb{C}}) \oplus (R(q)_{\mathbb{C}} \otimes \mathcal{H}^\perp)$$

It follows that p is a Hopf algebra morphism.

(b) The expression for $g^{\lambda}_{\mu\nu}$ can be derived directly from definitions (see 9.1, 8.2). The coefficients $g^{\mu\nu}_{\lambda}$ also can be computed directly, but the easier way is to use the self-adjointness of $R(q)$. Consider the inner product $\langle\, , \rangle_q$ on $\mathcal{H}$, the restriction of $\langle\, , \rangle$ from $R(q) \otimes \mathbb{C}$. Clearly, the basis $\{|Z(K_\lambda)| \cdot \chi_\lambda \mid \lambda \in \mathcal{P}\}$ in $\mathcal{H}$ is dual to $\{\chi_\lambda\}$ with respect to $\langle\, , \rangle_q$. So, by the axiom (S) from I.4 the matrix of m^* in the basis $\{|Z(K_\lambda)| \cdot \chi_\lambda\}$ is transposed of the matrix of m in the basis $\{\chi_\lambda\}$. This proves the desired formula for $g^{\mu\nu}_{\lambda}$

Q.E.D.

10.2. <u>Remark</u>. The coefficients $g^{\lambda}_{\mu\nu}$ are called <u>Hall polynomials</u> (they are polynomials in q, the number of elements of the basic field). The constants $|Z(K_\lambda)|$ are computed in [15]. Let us evaluate $|Z(K_{(n)})|$, which will be needed later. Choose an element $u \in K_{(n)}$ i.e. a Jordan cell of order n with the eigenvalue 1. Clearly, $|Z(K_{(n)})|$ equals $|Z_{G_n}(u)|$, the number of elements of the centralizer of u in G_n. It is easy to see that $Z_{G_n}(u)$ consists of elements of the form $F(u-1)$, where F is a polynomial over $\mathbb{F}_q$ of degree at most $n-1$ with non-zero constant term. It follows that

$$|Z(K_{(n)})| = q^{n-1}(q-1).$$

10.3. Now we show that the Hall algebra $\mathcal{H}$ is isomorphic as a Hopf algebra to $R_{\mathbb{C}} = R \otimes \mathbb{C}$, where R is the universal PSH-algebra (see §§ 3,4). Define the Hopf algebra morphism

$$p_\iota : R_{\mathbb{C}} \longrightarrow \mathcal{H}$$

to be the composition

$$R_{\mathbb{C}} \xrightarrow{\varphi_{\iota}} R(q)_{\mathbb{C}} \xrightarrow{p} \mathcal{H} \qquad \text{(see 9.6)}$$

<u>Theorem</u>. (a) The morphism $p_{\iota} : R_{\mathbb{C}} \longrightarrow \mathcal{H}$ is an isomorphism of Hopf algebras.

(b) For $n \geqslant I$ we have

$$p_{\iota}(x_n) = \sum_{\lambda \in \mathcal{P}_n} \chi_{\lambda} \; ;$$

$$p_{\iota}(y_n) = q^{\frac{n(n-I)}{2}} \cdot \chi_{(I^n)} \; ;$$

$$p_{\iota}(z_n) = \sum_{\lambda \in \mathcal{P}_n} (I-q)(I-q^2)\ldots(I-q^{r(\lambda)-I}) \cdot \chi_{\lambda} \qquad \text{(see 3.7)}$$

Theorem is proved in I0.4-I0.6.

I0.4. First we show that I0.3 (a) follows from the fact that all $p_{\iota}(z_n)$ are non-zero for $n \geqslant I$; in particular, it follows from I0.3 (b).

Indeed, since $\mathcal{H}$ is an epimorphic image of $R(q)_{\mathbb{C}}$, the multiplication and comultiplication in $\mathcal{H}$ are commutative. By Theorem quoted in 2.I, $\mathcal{H}$ is the symmetric algebra of the space of its primitive elements (see Appendix I, AI.I and AI.2 (a)). Since $R \otimes \mathbb{C} = \mathbb{C}[z_I, z_2, \ldots]$ (see 3.I5 (b)) and each $p_{\iota}(z_n)$ is primitive, the assumption that all $p_{\iota}(z_n)$ are non-zero, implies that p_{ι} is an embedding. Since

$$\dim \mathcal{H}_n = \dim (R \otimes \mathbb{C})_n = |\mathcal{P}_n|$$

it follows that p_{ι} is an isomorphism, as desired.

I0.5. The identity

$$p_{\iota}(x_n) = \sum_{\lambda \in \mathcal{P}_n} \chi_{\lambda}$$

follows at once from the fact that $\varphi_\iota(x_n)$ is the identity representation of G_n (see 9.6). To prove two other identities from I0.3 (b) we recall some well-known combinatorial results.

For $n \geq 0$ put

$$\varphi_q(n) = (q-1)(q^2-1)\dots(q^n-1).$$

For $0 \leq k \leq n$ denote by

$$\binom{n}{k}_q$$

the number of k-dimensional subspaces in a n-dimensional vector space over $\mathbb{F}_q$ (we put also $\binom{n}{k}_q = 0$ if the condition $0 \leq k \leq n$ does not hold). It is easy to see that the number of (ordered) k-frames in $\mathbb{F}_q^n$ equals

$$(q^n-1)(q^n-q)\dots(q^n-q^{k-1}) = q^{\frac{k(k-1)}{2}} \cdot \varphi_q(n)/\varphi_q(n-k) .$$

In particular, we have

$$|G_n| = q^{\frac{n(n-1)}{2}} \cdot \varphi_q(n) ;$$

$$\binom{n}{k}_q = \frac{\varphi_q(n)}{\varphi_q(k) \cdot \varphi_q(n-k)}$$

From the latter identity (or, directly, by definition of $\binom{n}{k}_q$) one can derive the relations

$$\binom{n}{k}_q = q^k \cdot \binom{n-1}{k}_q + \binom{n-1}{k-1}_q = \binom{n-1}{k}_q + q^{n-k} \cdot \binom{n-1}{k-1}_q$$

These relations in turn imply (by induction on n) the "q-binomial formula" :

$$(a+b)(a+qb)(a+q^2b)\cdot\ldots\cdot(a+q^{n-1}b) = \sum_{k=0}^{n} q^{\frac{k(k-1)}{2}} \binom{n}{k}_q \cdot a^{n-k}b^k$$

10.6. Denote temporarily by $\widetilde{x}_n$, $\widetilde{y}_n$, and $\widetilde{z}_n$ the right-hand sides of the identities 10.3 (b), and consider the formal power series

$$\widetilde{X}(\xi) = \sum_{n\geq 0} \widetilde{x}_n \cdot \xi^n \quad , \quad \widetilde{Y}(\xi) = \sum_{n\geq 0} \widetilde{y}_n \cdot \xi^n \quad \text{and}$$

$$\widetilde{Z}(\xi) = \sum_{n\geq 0} \widetilde{z}_{n+1} \cdot \xi^n$$

in $\mathcal{H}[[\xi]]$. By 3.16, to prove the identities

$$p_\iota(y_n) = \widetilde{y}_n \quad \text{and} \quad p_\iota(z_n) = \widetilde{z}_n$$

it suffices to verify that

$$\widetilde{X}(\xi)\,\widetilde{Y}(-\xi) = 1 \quad \text{and} \quad \left(\frac{d}{d\xi}\widetilde{X}(\xi)\right)\cdot \widetilde{Y}(-\xi) = \widetilde{Z}(\xi).$$

Expanding the products, we see that we must prove the identities

$$(*) \quad \sum_{k=0}^{n} (-1)^k\, \widetilde{y}_k\, \widetilde{x}_{n-k} = 0 \quad \text{and} \quad \sum_{k=0}^{n} (-1)^k (n-k)\widetilde{y}_k \widetilde{x}_{n-k} = \widetilde{z}_n$$

$(n\geq 1)$.

One can easily compute the products $\widetilde{y}_k\cdot\widetilde{x}_{n-k}$ by means of 10.1 (b). We obtain:

$$\widetilde{y}_k\cdot\widetilde{x}_{n-k} = \sum_{|\lambda|=n} q^{\frac{k(k-1)}{2}} \cdot \binom{r(\lambda)}{k}_q \cdot \chi_\lambda$$

(we use that for $u\in K_\lambda$ $\dim \operatorname{Ker}(u-1) = r(\lambda)$). Substituting these expressions into $(*)$, we see that it suffices to verify the identities

$$\sum_{k\geq 0} (-1)^k\, q^{\frac{k(k-1)}{2}} \binom{r}{k}_q = 0 \qquad (r\geq 1)\ ;$$

$$\sum_{k\geq 0} (-1)^k (r-k)\, q^{\frac{k(k-1)}{2}} \binom{r}{k}_q = (1-q)(1-q^2)\,\ldots\,(1-q^{r-1}) \qquad (r\geq 1).$$

The first identity is a special case of the q-binomial formula for a=I, b=-I, and n=r; to prove the second one it suffices to differentiate the q-binomial formula with respect to a and then to substitute again these values of a,b, and n.

This completes the proof of Theorem I0.3.

Q.E.D.

I0.7. We shall identify $\mathcal{H}$ with $R_{\mathbb{C}}$ via the isomorphism p_ι and denote identically an element of $R_{\mathbb{C}}$ and its image in $\mathcal{H}$. All structures on R and results of §§3,4 can be transferred to $\mathcal{H}$; for example,

$$\mathcal{H} = \mathbb{C}[x_I,x_2,\ldots] = \mathbb{C}[y_I,y_2,\ldots] = \mathbb{C}[z_I,z_2,\ldots],$$

On the other hand, we have another structures on $\mathcal{H}$, namely the inner product $\langle\, ,\rangle_q$ induced by the embedding $\mathcal{H} \hookrightarrow R(q)_{\mathbb{C}}$, and the basis $\{\chi_\lambda\}$, orthogonal with respect to $\langle\, ,\rangle_q$ (the inner product on $\mathcal{H}$, transferred from $R_{\mathbb{C}}$, will be denoted simply $\langle\, ,\rangle$).

The interaction of these structures gives rise to many interesting relations. Let us discuss some of them.

I0.8. Since $\mathcal{H}$ is identified with $R_{\mathbb{C}}$, the form δ_x (see 3.6). induces the C-linear form

$$\mathcal{H} \longrightarrow \mathbb{C},$$

which will be denoted also by δ_x.

<u>Proposition.</u> δ_x acts on $\mathcal{H}_n$ as follows:

$$\delta_x(\chi_\lambda) = 0 \quad \text{for } \lambda \neq (n), \text{ while } \delta_x(\chi_{(n)}) = I.$$

<u>Proof</u>. Define the form $\delta : \mathcal{H} \longrightarrow \mathbb{C}$ by

$$\begin{cases} \delta(\chi_\lambda) = 0 & \text{for } \lambda \neq \emptyset, (I),(2),\ldots \\ \delta(\chi_{(n)}) = I & \text{for } n \geq 0. \end{cases}$$

The expression for $g^{\lambda}_{\mu\nu}$ from 10.1 (b) implies that δ is multiplicative. Since the value of δ at any $x_n = \sum_{\lambda \in \mathcal{P}_n} \chi_\lambda$ equals 1, we have $\delta = \delta_x$.

Q.E.D.

This proposition can be reformulated as follows: for any irreducible representation $\omega \in R(L)_n$ (see 9.6) except the identity one, the character value of ω at the class $K_{(n)}$ is 0. In 11.10 we shall generalize this assertion.

As to the form δ_y, by definition (see 8.2, 9.5) it acts on $\mathcal{H}_n$ by

$$\delta_y(\chi_\lambda) = \frac{1}{|U_n|} \sum_{u \in K_\lambda \cap U_n} \psi^{-1}(u) .$$

It would be interesting to calculate the right-hand side more explicitly.

10.9. Now we compute the inner product $\langle\ ,\ \rangle_q$ between elements of various bases of $\mathcal{H}$. Note that $\mathcal{H}$ satisfies the axiom (S) from 1.4 with respect to each of inner products $\langle\ ,\ \rangle$ and $\langle\ ,\ \rangle_q$. It follows that all arguments of Chapter 1, which are not based on the axiom (P), make sense if one replaces $\langle\ ,\ \rangle$ by $\langle\ ,\ \rangle_q$. In particular, all assertions in 1.9 except 1.9 (d) remain valid for operators $\mathcal{H} \to \mathcal{H}$, adjoint to the operators of multiplication with respect to the inner product $\langle\ ,\ \rangle_q$.

<u>Proposition</u>. (a). The operator $t : \mathcal{H} \to \mathcal{H}$ obtained by extending $t : R \longrightarrow R$ by $\mathbb{C}$ - linearity (see 3.11, 10.8) preserves the inner product $\langle\ ,\ \rangle_q$.

(b) We have

$$\langle x_n, x_n \rangle_q = \langle y_n, y_n \rangle_q = \frac{q^{\frac{n(n-1)}{2}}}{\varphi_q(n)}$$

$$\langle x_n, y_n\rangle_q = \frac{I}{\varphi_q(n)} ;$$

$$\langle x_n, z_n\rangle_q = \frac{I}{q^n - I} ;$$

$$\langle y_n, z_n\rangle_q = (-I)^{n-I} \cdot \frac{I}{q^n - I} ;$$

$$\langle z_n, z_n\rangle_q = \frac{n}{q^n - I} .$$

(c) For any $\lambda = (l_I, \ldots, l_r) \in \mathcal{P}$ with $l_i \neq 0$ put

$$\Pi_q(\lambda) = \prod_{i=I}^{r} (q^{l_i} - I)$$

Then for any $x \in \mathcal{H} = R_{\mathbb{C}}$ we have

$$\langle x, z_\lambda\rangle_q = \frac{I}{\Pi_q(\lambda)} \cdot \langle x, z_\lambda\rangle ,$$

Combining this equality with 3.I7 (c) and 4.I7 (b), we obtain the values of $\langle x_\mu, z_\lambda\rangle_q$, $\langle y_\mu, z_\lambda\rangle_q$, $\langle z_\mu, z_\lambda\rangle_q$ and $\langle \{\varkappa\}, z_\lambda\rangle_q$.

(d) For $\lambda, \mu \in \mathcal{P}$ we have

$$\langle x_\lambda, x_\mu\rangle_q = \langle y_\lambda, y_\mu\rangle_q = \sum_{(k_{ij}) \in M^{\lambda,\mu}} \prod_{i,j} \frac{q^{\frac{k_{ij}(k_{ij}-I)}{2}}}{\varphi_q(k_{ij})} ;$$

$$\langle x_\lambda, y_\mu\rangle_q = \sum_{(k_{ij}) \in M^{\lambda,\mu}} \prod_{i,j} \frac{I}{\varphi_q(k_{ij})}$$

(the set $M^{\lambda,\mu}$ is defined in 3.I7 (c)).

<u>Proof</u>. (a). The proof is quite similar to that of 3.II (a).

(b) The expressions for $\langle x_n, y_n\rangle_q$, $\langle y_n, y_n\rangle_q$, and $\langle z_n, y_n\rangle_q$ follow at once from definition of $\langle\,,\rangle_q$ (see 8.2) with the account of I0.3 (b) and I0.5. By (a),

$$\langle x_n, x_n\rangle_q = \langle y_n, y_n\rangle_q \quad \text{and} \quad \langle z_n, x_n\rangle_q = (-I)^{n-I} \cdot \langle z_n, y_n\rangle_q$$

(see 3.II (b), 3.I7 (a)). Finally, the equality

$$\langle z_n, z_n\rangle_q = \frac{n}{q^n - I}$$

is a special case of (c).

(c). By means of the axiom (S), we reduce our equality to the case when $\lambda = (n)$ and x runs over some basis in $\mathcal{H}_n$. By I.7, for $\mu \neq (n)$ we have

$$\langle z_n, x_\mu\rangle = \langle z_n, x_\mu\rangle_q = 0;$$

it follows from (b) and 3.I5 (a) that

$$\langle z_n, x_n\rangle_q = \frac{I}{q^n - I} \cdot \langle z_n, x_n\rangle,$$

as desired.

(d) The proof is quite similar to that of 3.I7 (c).

Q.E.D.

I0.I0. <u>Corollary</u>. The number of unipotent elements in G_n equals $q^{n(n-I)}$.

<u>Proof</u>. By definition of $\langle\,,\rangle_q$ and I0.3 (b) the number of unipotent elements in G_n equals

$$|G_n| \cdot \langle x_n, x_n\rangle_q$$

Now apply I0.9 (b) and I0.5.

Q.E.D.

This is a special case of the well-known theorem by R.Steinberg.

10.11. Now consider the elements $\{\lambda\}\in\mathcal{H}$ (see 10.7). Since we know the decomposition of $\{\lambda\}$ in each of the bases (x_μ), (y_μ), and (z_μ) (see 4.16, 4.17), Proposition 10.9 in principle enables us to compute the inner products $<\{\lambda\},x_\mu>_q$, $<\{\lambda\},y_\mu>_q$, $<\{\lambda\},z_\mu>_q$ and $<\{\lambda\},\{\mu\}>_q$. For a general μ , the author does not know simple formulas for these expressions (except $<\{\lambda\},z_\mu>_q$; see 10.9 (c)). Let us compute $<\{\lambda\},y_n>_q$ (in 11.10 this will be applied for the computation of dimensions of all irreducible representations of G_n). Denote by d_q the linear form $\mathcal{H}\longrightarrow\mathbb{C}$, acting on each $\mathcal{H}_n$ by

$$d_q(x) = <x,y_n>_q$$

<u>Proposition</u>. (a) The form $d_q:\mathcal{H}\longrightarrow\mathbb{C}$ is multiplicative, (b) For any $\lambda\in\mathcal{P}$ with c.f. $(\lambda) = (l_1,l_2,\dots)$ (see 3.7) we have

$$d_q(\{\lambda\}) = \frac{q^{\sum_{k\geq 1}(k-1)l_k}}{\prod\limits_{a\in\lambda}(q^{h(a)}-1)}$$

(here $h(a)$ is hook length of a; see 6.5 (b)).

<u>Proof</u>. (a) The proof is quite similar to that in 3.6.

(b) Under the notation of 4.16 (a) we have

$$\{\lambda\} = \det\left(x_{L_i-r+j}\right)_{i,j=1,\dots,r}$$

Applying the multiplicative form d_q to this determinant, we obtain that

$$d_q(\{\lambda\}) = \det\left(\frac{1}{\varphi_q(L_i-r+j)}\right).$$

(see 10.9 (b)). The computation of the determinant is quite similar to that in 6.5 (the details are left to the reader).

Q.E.D.

10.12. One can use the information on $\mathcal{H}$ to obtain a lot of interesting identities. Let us consider one example. By 3.16 we have

$$Z(-\xi) = \frac{d}{d\xi} \ln Y(\xi)$$

Let us apply to this identity the multiplicative form $d_q : R \longrightarrow \mathbb{Q}$ coefficientwise (see 3.16 (*)) and use 10.9 (b). We obtain the identity in $\mathbb{Q}[[\xi]]$ which can be rewritten as

$$\exp\Big(\sum_{k\geq 1} \frac{\xi^k}{k(q^k-1)}\Big) = \sum_{n\geq 0} \frac{q^{\frac{n(n-1)}{2}}}{\varphi_q(n)} \xi^n$$

10.13. Now we consider the relationships between the basis $\{\chi_\lambda\}$ in $\mathcal{H}$ and various bases in R (see §§3,4). Let $(Q(\lambda,\mu))$ be the transition matrix between the bases (χ_λ) and (z_μ), i.e.

$$(*) \qquad z_\mu = \sum_\lambda Q(\lambda,\mu)\cdot\chi_\lambda \qquad (\lambda,\mu\in\mathcal{P})$$

The coefficients $Q(\lambda,\mu)$ are called <u>Green polynomials</u> (they happen to be polynomials in q); they play the important role in the character theory of the groups G_n (see [4] and 11.8 below).

By 10.3 (b), for $\lambda\in\mathcal{P}_n$ we have

$$Q(\lambda,(n)) = (1-q)(1-q^2)\dots(1-q^{r(\lambda)-1});$$

obviously, all $Q(\lambda,\mu)$ can be expressed in terms of these values and the structural constants $g^\lambda_{\mu\nu}$. In particular, $Q(\lambda,\mu)\in\mathbb{Z}$.

Each of the bases (χ_λ) and (z_λ) is orthogonal with respect to $\langle\,,\,\rangle_q$, and we have

$$\langle \chi_\lambda, \chi_\lambda \rangle_q = \frac{1}{|Z(K_\lambda)|}, \qquad \langle z_\lambda, z_\lambda \rangle_q = \frac{c_\lambda}{\Pi_q(\lambda)}$$

(see 10.1, 10.9 (c), and 3.18). So we have the inversion of (*):

$$(**) \qquad \chi_\lambda = \sum_\mu \frac{\Pi_q(\mu)}{c_\mu \cdot |Z(K_\lambda)|} \cdot Q(\lambda, \mu)\, z_\mu$$

Combining (*) and (**) with the results of §§ 3,4, we obtain the transition matrices between (χ_λ) and all bases from §§ 3,4 in terms of Green polynomials.

The main properties of $Q(\lambda, \mu)$ are given in [15] ,Prop. 5.16; all of them become very transparent via our approach to the Hall algebra $\mathcal{H}$. For example, comparing (*) and (**), we obtain the <u>orthogonality relations</u>:

$$\sum_\lambda \frac{1}{|Z(K_\lambda)|} Q(\lambda, \mu) \cdot Q(\lambda, \nu) = \frac{c_\nu}{\Pi_q(\nu)} \cdot \delta_{\mu\nu};$$

$$\sum_\mu \frac{\Pi_q(\mu)}{c_\mu} \cdot Q(\lambda, \mu) \cdot Q(\nu, \mu) = |Z(K_\lambda)| \cdot \delta_{\lambda\nu}.$$

We shall give two more identities on $Q(\lambda, \mu)$. Let us decompose x_n with respect to the basis (χ_λ) by two different ways. By 10.3 (b),

$$x_n = \sum_{\lambda \in \mathcal{P}_n} \chi_\lambda .$$

On the other hand, by 3.17 (b), 3.18, and (*) we have

$$x_n = \sum_{\mu \in \mathcal{P}_n} z_\mu^{\perp} = \sum_{\mu \in \mathcal{P}_n} c_\mu^{-1} \cdot z_\mu = \sum_{\lambda \in \mathcal{P}_n} \chi_\lambda \cdot \sum_{\mu \in \mathcal{P}_n} c_\mu^{-1} \cdot Q(\lambda, \mu)$$

Comparing these expressions, we obtain the identity

$$\sum_{\mu} c_{\mu}^{-1} \cdot Q(\lambda,\mu) = 1 \quad \text{for all } \lambda \in \mathcal{P}.$$

Similarly, using y_n instead of x_n, we obtain the identity

$$\sum_{\mu \in \mathcal{P}_n} (-1)^{n-r(\mu)} \cdot c_{\mu}^{-1} \cdot Q(\lambda,\mu) = q^{\frac{n(n-1)}{2}} \cdot \delta_{\lambda,(1^n)} .$$

§ 11. The character values of $GL(n, F_q)$ at unipotent elements

11.1. In this section we compute the character values of all irreducible representations of the groups G_n at unipotent elements. Clearly, this problem is equivalent to that of explicit computation of the morphism

$$p : R(q) \longrightarrow \mathcal{H} \qquad \text{(see 10.1)}$$

We recall that $R(q) = \bigotimes_{\rho \in \mathcal{C}} R(\rho)$, and each $R(\rho)$ is identified with R via the isomorphism φ_ρ (see 9.3-9.5); on the other hand, $\mathcal{H}$ is identified with $R_{\mathbb{C}}$ via the isomorphism p_ι (see 10.3). Since p is a ring homomorphism (see 10.1 (a)), to compute it explicitly one has only to compute for each $\rho \in \mathcal{C}$ the composition

$$R \xrightarrow{\varphi_\rho} R(\rho) \xrightarrow{p} \mathcal{H} \xrightarrow{p_\iota^{-1}} R_{\mathbb{C}}$$

Denote this composition by $p_\rho : R \longrightarrow R_{\mathbb{C}}$. Clearly, p_ρ is a Hopf algebra morphism, but it does not preserve the grading; if $\deg \rho = k$ i.e. $\rho \in \Omega(G_k)$ then p_ρ maps R_n to $(R_{\mathbb{C}})_{kn}$.

11.2. Our definition of φ_ρ is based on the identity $\delta(\rho) = 1$ from 9.4. Now we prove this identity. The present proof is due to J.N.Bernstein; it is based on the technique developed in § IO. Another proof of the inequality $\delta(\rho)>0$ (i.e. of the S.I.Gelfand theorem) will be given in I3.4. The proof finishes in II.6.

Since $\mathcal{H}$ is identified with $R_{\mathbb{C}}$, the form δ_x (δ_y) from 3.6 induces the $\mathbb{C}$-linear multiplicative form $\mathcal{H}\to\mathbb{C}$, which will be denoted also by δ_x (δ_y)

<u>Lemma</u>. The form δ on R(q) (see 9.4) is equal to the composition

$$R(q)\xrightarrow{\ p\ }\mathcal{H}\xrightarrow{\ \delta_y\ }\mathbb{C}$$

Proof. Remembering the definition of φ_i (see 9.6), we see that Lemma is equivalent to the fact that δ vanishes on the orthogonal complement $\mathcal{H}^{\perp}$ of $\mathcal{H}$ in $R(q)_{\mathbb{C}}$, i.e. on the space of functions, taking value 0 at unipotent elements. This fact follows at once from definition of δ and 8.2.

Q.E.D.

II.3. Denote by $\delta' : R(q)\longrightarrow\mathbb{C}$ the composition

$$R(q)\xrightarrow{\ p\ }\mathcal{H}\xrightarrow{\ \delta_x\ }\mathbb{C}$$

By IO.8, for every $\pi\in\mathcal{A}(G_n)$ the value $\delta'(\pi)$ is the character value of π at the class $K_{(n)}$.

Lemma. (a) There exists an involution

$$\omega\longmapsto\omega^t$$

of the set $\Omega(G_n)$ such that $\rho^t=\rho$ for every cuspidal ρ, and $\delta(\omega)=\pm\delta'(\omega^t)$ for every $\omega\in\Omega(G_n)$.

(b) For $n\geqslant I$ we have

$$\sum_{\omega\in\Pi(G_n)}\delta(\omega)^2=q^{n-I}(q-I)\,.$$

<u>Proof</u>. (a) For each $\rho\in\mathcal{C}$ consider the unique non-trivial automorphism t of the PSH-algebra $R(\rho)$ (see 3.I (f)).

We extend these automorphisms to the automorphism $\omega \mapsto \omega^t$ of the PSH-algebra $R(q) = \bigotimes_{\rho \in \mathcal{C}} R(\rho)$; clearly, $\rho^t = \rho$ for any $\rho \in \mathcal{C}$. By definition of t (see 3.II), for any $\omega \in \Omega(G_n)$ we have

$$\omega^t = \pm T(\omega),$$

where $T : R(q) \longrightarrow R(q)$ is the conjugation of $R(q)$. On the other hand, by definition of δ_x and δ_y, for any $\pi \in \mathcal{H}_n$ we have

$$\delta_y(\pi) = \pm \delta_x \circ T_{\mathcal{H}}(\pi),$$

where $T_{\mathcal{H}}: \mathcal{H} \longrightarrow \mathcal{H}$ is the conjugation of $\mathcal{H}$. By definition (see Appendix I, AI.6), T commutes with any Hopf algebra morphism; in particular,

$$T_{\mathcal{H}} \circ p = p \circ T : R(q) \longrightarrow \mathcal{H}$$

(see 10.I (a)). Summarizing and applying Lemma II.2, we see that

$$\delta'(\omega^t) = \pm \delta_x \circ p \circ T(\omega) = \pm \delta_x \circ T_{\mathcal{H}} \circ p(\omega) = \pm \delta_y \circ p(\omega) = \pm \delta(\omega)$$

for all $\omega \in \Omega(G_n)$, as desired.

(b) By (a), it suffices to verify that

$$(*) \quad \sum_{\omega \in \Omega(G_n)} \delta'(\omega)^2 = q^{n-1}(q-1).$$

Set

$$\zeta_n = |Z(K_{(n)})| \cdot \chi_{(n)} \in \mathcal{H}_n \quad \text{(see 10.I)}$$

By definition,

$$\delta'(\omega) = \langle \zeta_n, \omega \rangle_q .$$

Therefore, the left-hand side of (*) equals

$$\sum_{\omega \in \Omega(G_n)} \langle \zeta_n, \omega \rangle_q^2 = \langle \zeta_n, \zeta_n \rangle_q = |Z(K_{(n)})| = q^{n-1}(q-1)$$

as desired (see I0.2).

Q.E.D.

II.4. Now we are able to prove the S.I.Gelfand theorem, i.e. that $\delta(\rho) \neq 0$ for any $\rho \in \mathcal{C}$. By II.3 (a) it suffices to prove that $\delta'(\rho) \neq 0$, i.e. that $\delta_x(p(\rho)) \neq 0$. Let $\deg \rho = k$. Since $p : R(q) \longrightarrow \mathcal{H}$ is a Hopf algebra morphism, the element $p(\rho) \in \mathcal{H}_k$ is primitive. Therefore, $p(\rho)$ is proportional to z_k. But $\delta_x(z_k) = I$ (see 3.I5), so $\delta_x(p(\rho))$ can be zero only if $p(\rho) = 0$. The equality $p(\rho) = 0$ means that the character of ρ takes value 0 at all unipotent elements in G_k; in particular, $\dim \rho = 0$. This contradiction shows that $\delta'(\rho) \neq 0$, as desired.

II.5. To prove the Gelfand -Graev theorem, i.e. that $\delta(\omega) \leq I$ for any $\omega \in \Omega(G_n)$, we need the following well-known.

<u>Proposition</u>. For each $n \geq I$ the number of cuspidal representations of G_n equals the number of irreducible polynomials $P \in \mathbb{F}_q[T]$ of degree n with leading coefficient I and non-zero constant term.

<u>Proof</u>. First we compute the number of conjugacy classes in G_n. It is well-known that these classes are in a natural one-to-one correspondence with isomorphism classes of $\mathbb{F}_q[T]$ - modules V such that $\dim_{\mathbb{F}_q} V = n$ and the operator $T : V \longrightarrow V$ is invertible. The structural theory of modules over principal ideal rings implies that each such module V decomposes into the direct sum of primary cyclic ones, and the isomorphism class of V is determined by this decomposition. Any primary cyclic module over $\mathbb{F}_q[T]$ has the form

$$\mathbb{F}_q[T] / (P^l) ,$$

where P is an irreducible polynomial over $\mathbb{F}_q$ (one can assume its leading coefficient to be I) and $l \geq 0$.

Summarizing, we obtain the following description of conjugacy classes in G_n. Denote by $\mathcal{C}'$ the graded set of irreducible polynomials $P \in \mathbb{F}_q[T]$ with leading coefficient I and non-zero constant term (see 4.I9). The conjugacy classes in G_n are parametrized by the set $S_n(\mathcal{C}'; \mathcal{P})$: to a function

$$\varphi : P \longmapsto (l_I(P), l_2(P), \ldots)$$

there corresponds the isomorphism class of the $\mathbb{F}_q[T]$ - module

$$\bigoplus_{P \in \mathcal{C}'} \bigoplus_{k \geq 1} \mathbb{F}_q[T] / (P^{l_k(P)}).$$

For example, the unipotent class K_λ from IO.I corresponds to the function φ such that $\varphi(T-I) = \lambda$, and $\varphi(P) = \emptyset$ for $P \neq T - I$.

Since the number of irreducible representations of G_n equals its number of conjugacy classes, it follows that

$$|S_n(\mathcal{C}; \mathcal{P})| = |S_n(\mathcal{C}'; \mathcal{P})| \qquad \text{for } n \geq I$$

(see 9.4; note that we don't use the explicit form of the isomorphisms Φ_ρ but only their existence). From this we easily derive by induction on n that for every $n \geq I$ the sets $\mathcal{C}$ and $\mathcal{C}'$ have the same number of elements of degree n.

Q.E.D.

II.6. <u>Proof of the Gelfand-Graev theorem</u>. The idea of the proof is to give a lower bound for the sum

$$S = \sum_{\omega \in \Omega(G_n)} \delta(\omega)^2 ;$$

we shall show that the assumption that $\delta(\omega) > I$ for some

$\omega \in \Omega(G_n)$, implies that

$$S > q^{n-I}(q-I).$$

This contradicts II.3 (b).

We shall use the notation of 2.5. By 2.5 (a),

$$\Omega(G_n) = \coprod_{\varphi \in S_n(\mathscr{C};\mathbb{Z}^+)} \Omega(\varphi) \quad ,$$

where

$$S_n(\mathscr{C};\mathbb{Z}^+) = \{\varphi \in S(\mathscr{C};\mathbb{Z}^+) \mid \deg\varphi = \sum_{\rho\in\mathscr{C}} \varphi(\rho)\cdot\deg\rho = n\}$$

Let us prove that

$$|S_n(\mathscr{C};\mathbb{Z}^+)| = q^{n-I}(q-I) .$$

Indeed, by II.5 we may replace the set $\mathscr{C}$ by $\mathscr{C}'$. Associating to any $\varphi \in S(\mathscr{C}';\mathbb{Z}^+)$ the polynomial

$$\prod_{P\in\mathscr{C}'} P^{\varphi(P)} ,$$

we obtain the bijection of $S_n(\mathscr{C}';\mathbb{Z}^+)$ with the set of polynomials $F \in \mathbb{F}_q[T]$ of degree n with leading coefficient I and non-zero constant term. The number of such polynomials equals $q^{n-I}(q-I)$, as desired.

For any $\varphi \in S(\mathscr{C};\mathbb{Z}^+)$ put

$$S(\varphi) = \sum_{\omega\in\Omega(\varphi)} \delta(\omega)^2 .$$

By definition, an element of $\Omega(\varphi)$ is an irreducible constituent of

$$\pi_\varphi = \prod_{\rho\in\mathscr{C}} \rho^{\varphi(\rho)} .$$

We have already proven that $\delta(\rho) > 0$ for all $\rho \in \mathscr{C}$ (see II.4). Since δ is multiplicative, we have $\delta(\pi_\varphi) > 0$. Since

δ is a positive form on R(q) with integer values, it follows that for any φ there exists $\omega \in \Omega(\varphi)$ such that $\delta(\omega) \geqslant I$. Therefore,

$$S(\varphi) \geqslant I \quad \text{for all } \varphi \in S(\mathcal{C}; \mathbb{Z}^+).$$

It follows that

$$S = \sum_{\deg \varphi = n} S(\varphi) \geqslant \sum_{\deg \varphi = n} I = q^{n-I}(q-I) ;$$

moreover, if $\delta(\omega) > I$ for some $\omega \in \Omega(G_n)$ then this inequality becomes strict. It remains to apply II.3 (b).

Q.E.D.

II.7. Now we compute the morphism

$$p_\rho : R \longrightarrow R_{\mathbb{C}} \qquad \text{(see II.I).}$$

Theorem. Let $\rho \in \mathcal{C}$ and $\deg \rho = k$. For $n \geqslant I$ we have

$$p_\rho(z_n) = (-I)^{n(k-I)} z_{kn} .$$

<u>Proof</u>. Since z_n is primitive and p_ρ is a Hopf algebra morphism, it follows that $p_\rho(z_n)$ is a primitive element of $(R_{\mathbb{C}})_{kn}$, hence it is propertional to z_{kn} (see 3.I5 (a)). To find the coefficient of proportionality we use Lemma II.2. We have

$$\delta_y(p_\rho(z_n)) = \delta(\varphi_\rho(z_n)) = \delta_y(z_n) = (-I)^{n-I} ,$$

while

$$\delta_y(z_{kn}) = (-I)^{kn-I}$$

(see 3.I7 (b)). Hence the coefficient equals $(-I)^{kn-n} = (-I)^{n(k-1)}$

Q.E.D.

II.8. Theorem II.7 implies at once the J.Green formula for character values of irreducible representation of G_n at unipotent elements. For any $\mu = (m_1, \ldots, m_r) \in \mathcal{P}$ and $k \in \mathbb{N}$ denote by $k \cdot \mu$ the partition $(km_1, km_2, \ldots, km_r) \in \mathcal{P}$; for any finite family $(\mu_\alpha \mid \alpha \in A)$ of partitions denote by $\coprod_{\alpha \in A} \mu_\alpha$ the union of these partitions (for example,

$$(4,2,1) \coprod (3^2,2,1) = (4,3^2,2^2,1^2) \quad).$$

The J.Green formula says that the character value of the irreducible representation $\{\varphi\}$ of G_n (here $\varphi \in S_n(\mathcal{C};\mathcal{P})$; see 9.4-9.5) at the class K_λ equals

$$(-1)^{n-\sum_\rho |\varphi(\rho)|} \cdot \sum_{\psi \in S(\mathcal{C};\mathcal{P})} \left[Q(\lambda, \coprod_{\rho \in \mathcal{C}} \deg\rho \cdot \psi(\rho)) \cdot \prod_{\rho \in \mathcal{C}} \frac{\langle \{\varphi(\rho)\}, z_{\psi(\rho)} \rangle}{c_{\psi(\rho)}} \right]$$

Here the $Q(\lambda, \mu)$ are J.Green polynomials (see 10.13); the inner products $\langle \{\nu\}, z_\mu \rangle$ are computed in 4.17 while the constants c_μ in 3.17 (c) (note that the values $\langle \{\nu\}, z_\mu \rangle$ are the character values of irreducible representations of symmetric groups; see 6.3).

To prove the J.Green formula it suffices to decompose the element

$$p(\{\varphi\}) = \prod_{\rho \in \mathcal{C}} p_\rho(\{\varphi(\rho)\})$$

with respect to the basis $\{z_\mu\}$ and then pass to the basis $\{\chi_\lambda\}$ of $\mathcal{H}$; the details are left to the reader.

Note that the very simple formula of Theorem 11.7 is essentially equivalent to the J.Green formula.

11.9. <u>Corollary</u>. Let $\rho \in \mathcal{C}$ and $\lambda \in \mathcal{P}$ be such that $\deg \rho = |\lambda| = n$. Then the character value of ρ at the class K_λ equals

$$(-1)^{n-r(\lambda)} \cdot \varphi_q(r(\lambda)-1) \, ;$$

in particular, this value does not depend on ρ.

This follows at once from II.7 and IO.3 (b).

II.IO. Now we apply II.7 to compute the character values of irreducible representations of G_n at the classes $K_{(I^n)}$ and $K_{(n)}$ (the character value of π at $K_{(I^n)}$ is, evidently, the dimension of π).

<u>Proposition</u>. Let $\varphi \in S_n(\mathcal{C};\mathcal{P})$ and $\{\varphi\}$ be the corresponding irreducible representation of G_n.

(a) $\dim\{\varphi\} = \varphi_q(n)\cdot \prod_{\rho\in\mathcal{C}} d_q \deg\rho(\{\varphi(\rho)\})$ (see IO.II)

(b) The character value of $\{\varphi\}$ at $K_{(n)}$ is 0 unless any diagram $\varphi(\rho)$ for $\rho\in\mathcal{C}$ has only one row; in this case this value equals

$$(-I)^{n-\sum_{\rho\in\mathcal{C}}|\varphi(\rho)|}$$

<u>Proof</u>. (a) Define the form $d : R(q) \longrightarrow \mathbb{C}$ to be the composition

$$R(q) \xrightarrow{\ D\ } \mathcal{H} \xrightarrow{\ d_q\ } \mathbb{C} \qquad \text{(see IO.II)}$$

Clearly,

$$d(\pi) = \frac{\dim \pi}{\varphi_q(n)} \qquad \text{for any } \pi\in \mathcal{A}(G_n).$$

By IO.I (a) and IO.II (a), d is multiplicative. Therefore, our assertion follows from the identity

$$(*) \qquad d\circ\varphi_\rho = d_q \deg\rho \quad : R_{\mathbb{Q}} \longrightarrow \mathbb{Q}$$

By IO.9 (b), d_q can be defined as a multiplicative form on $R_{\mathbb{Q}}$, whose value at z_n is $(-I)^{n-I}/q^n-I$ $(n\geqslant I)$. So (*) follows at once from II.7.

(b) The proof is quite similar to that of (a). One has only

to replace d by the form δ' (see II.3), and the identity (x) by the assertion that

$$\delta' \circ \varphi_\rho = (-1)^{(\deg\rho - 1)n} \cdot \delta_x \qquad \text{on } R_n .$$

Q.E.D.

II.11. We conclude this section with a very simple proof of the Macdonald conjecture (see [24], 6.11).

<u>Proposition</u>. For any $\omega \in \Omega(G_n)$ the sum of character values of ω over all unipotent elements of G_n equals $\pm q^m \cdot \dim \omega$ for some $m \in \mathbb{Z}$.

<u>Proof</u>. Under the notation of § 10, the sum in question equals $|G_n| \cdot \langle p(\omega), x_n \rangle_q$ (see 10.1. 10,3 and 10.7). We have

$$|G_n| \cdot \langle p(\omega), x_n \rangle_q = |G_n| \cdot \langle t(p(\omega)), y_n \rangle_q =$$

$$= \pm |G_n| \cdot \langle p(\omega^t), y_n \rangle_q = \pm q^{\frac{n(n-1)}{2}} \cdot \dim \omega^t$$

(see 10.9 (a), Proof of 11.3 (a) and 10.3 (b)). It remains to prove that $\dim \omega^t / \dim \omega$ is a power of q. Clearly, if $\omega = \{\varphi\}$ ($\varphi \in S_n(\mathcal{C}; \mathcal{P})$) then $\omega^t = \{\varphi^t\}$, where $\varphi^t(\rho) = (\varphi(\rho))^t$ for $\rho \in \mathcal{C}$. Our statement follows at once from 11.10 (a) and 10.11 (one has only to observe that the set of hook lengths of any Young diagram coincides with that of the transposed diagram).

Q.E.D.

11.12. <u>Remark</u>. In [26] C.W.Curtis defined for any finite Chevalley group G the <u>duality</u> operation $R(G) \longrightarrow R(G)$ $(\omega \mapsto \omega^*)$. The definition easily implies that

$$\omega^* = (-1)^n \cdot T(\omega) \qquad \text{for any} \qquad \omega \in R(G_n),$$

where T is the conjugation of R(q). In [26],[27] the main

properties of T, namely that T is an involutive Hopf algebra automorphism and an isometry, are generalized for all finite Chevalley groups. Using these properties and one result of T.A.Springer, D.Alvis in [27] proved the weakened form of the Macdonald conjecture for all finite Chevalley groups (the arguments in [27] are similar to those of II.II).

§ I2. Degenerate Gelfand-Graev modules

I2.I. By degenerate Gelfand-Graev modules we mean representations of G_n induced by various one-dimensional representations of the subgroup $U = U_n$ of unipotent upper triangular matrices. I.M.Gelfand and M.I.Graev in [I3] proved that any irreducible representation of G_n can be embedded into some of these modules. We obtain the more precise result, computing explicitly the spectrum of any of these modules.

For each ordered partition $(k_I,\dots,k_r)$ of n, we define the character $\psi_{k_I,\dots,k_r}$ of U by

$$\psi_{k_I,\dots,k_r}((u_{ij})) = \psi(\sum u_{i,i+I}),$$

the sum is over all i except $n-k_I$, $n-k_I-k_2,\dots,n-k_I-\dots-k_{r-I}$. For example, $\psi_n = \psi$ (see 9.4) while $\psi_{1,1,\dots,1} = I$. It is easy to see that every character of U is conjugate to one of the $\psi_{k_I,\dots,k_r}$ under the action of the diagonal subgroup in G_n. It follows that any degenerate Gelfand-Graev module is isomorphic to one of the modules

$$\pi_{k_I,\dots,k_r} = \mathrm{Ind}_{U_n}^{G_n} \psi_{k_I,\dots,k_r}$$

Theorem. Let $\varphi \in S_n(\mathcal{C};\mathcal{P})$ and $\{\varphi\}$ be the corresponding irreducible representation of G_n (see 9.4, 9.5). The multiplicity $\langle\{\varphi\}, \pi_{k_1,\ldots,k_r}\rangle$ of $\{\varphi\}$ in the module $\pi_{k_1,\ldots,k_r}$ is

$$\sum \prod_{\rho\in\mathcal{C}} \langle\{\varphi(\rho)\}, y_{(l_1(\rho), l_2(\rho),\ldots,l_r(\rho))}\rangle ,$$

where the sum is over all functions

$$\rho \longmapsto (l_I(\rho), l_2(\rho), \ldots, l_r(\rho)\ ) \quad \text{from } \mathcal{C} \quad \text{to } (\mathbb{Z}^+)^r$$

such that

$$\sum_{\rho\in\mathcal{C}} l_i(\rho)\cdot \deg\rho \ = k_i \qquad \text{for} \quad i = I,\ldots,r$$

(the inner products $\langle\{\mu\}, y_\lambda\rangle$, appearing in the answer, are computed in 4.I4).

Theorem is proved in I2.2-I2.3.

I2.2. By definition,

$$\pi_{k_I,\ldots,k_r} = i_{U,\psi_{k_I,\ldots,k_r}}(I),$$

where I stands for the identity representation of the identity group (see 8.I). According to 9.I (b), for any $\omega\in \mathcal{A}(G_n)$ we have

$$\langle\omega, \pi_{k_I,\ldots,k_r}\rangle = r_{U,\psi_{k_I,\ldots,k_r}}(\omega) .$$

We want to express $r_{U,\psi_{k_I,\ldots,k_r}}$ in terms of the PSH-algebra structure on $R(q)$. Define the operator $D : R(q) \longrightarrow R(q)$ to be the composition

$$R(q) \xrightarrow{m^*} R(q)\otimes R(q) \xrightarrow{id\otimes\delta} R(q)\otimes\mathbb{Z} \xrightarrow{\ \sim\ } R(q)$$

(see 9.I, 9.4); denote by D_k the homogeneous component of D, acting from $R(G_n)$ to $R(G_{n-k})$.

Lemma. The operator $r_{U,\psi\, k_1,\ldots,k_r} : R(G_n) \longrightarrow R(G_0)=\mathbb{Z}$ coincides with the composition

$$D_{k_r} \circ D_{k_{r-1}} \circ \ldots \circ D_{k_1}$$

Proof. Put

$$U'_k = \{ u = (u_{ij}) \in U_n \mid u_{ij} = \delta_{ij} \quad \text{for} \quad 1 \leq i \leq n-k \}$$

Remembering the definitions of m^* and δ (see 9.1,9.4) and using 8.1 (d), we see that

$$D_k : R(G_n) \longrightarrow R(G_{n-k})$$

equals the composition

$$r_{U'_k,\psi} \circ r_{U_{n-k,k},1} \quad .$$

To compute the composition $D_{k_r} \circ D_{k_{r-1}} \circ \ldots \circ D_{k_1}$ we apply several times Prop. 8.1 (c). The easy computation shows that this composition equals $r_{U,\psi\, k_1,\ldots,k_r}$.

Q.E.D.

12.3. According to 9.5 and 12.2,

$$\langle\{\varphi\},\pi_{k_1,\ldots,k_r}\rangle = D_{k_r} \circ \ldots \circ D_{k_1} \Big(\prod_{\rho\in\mathcal{C}} \varphi_\rho(\{\varphi(\rho)\}) \Big).$$

So Theorem 12.1 follows at once from the next.

Lemma. (a) $D_k \Big(\prod_{i=1}^{p} v_i \Big) = \sum D_{m_1}(v_1) \cdot D_{m_2}(v_2) \cdot \ldots \cdot D_{m_p}(v_p)$, where the sum is over all $(m_1,\ldots,m_p) \in (\mathbb{Z}^+)^p$ with $\sum m_i = k$.

(b) Each subspace $R(\rho)$ in $R(q)$ is invariant under D and hence under all D_m. If m is not divisible by $\deg\rho$ then D_m is 0 on $R(\rho)$; but if $m = l\cdot\deg\rho$, $l \in \mathbb{Z}^+$, then the operator

$$D_m \circ \varphi_\rho : R \longrightarrow R(\rho)$$

coincides with $\varphi_\rho \circ \gamma_l^*$.

<u>Proof</u>. Part (a) means that D is a ring homomorphism. This follows from the Hopf axiom (H) of $R(q)$ and the fact that δ is multiplicative (see 9.I, 9.4). Part (b) is an immediate consequence of definitions of D and φ_ρ (see 9.5).

Q.E.D.

I2.4. <u>Corollary.</u> If two sequences $(k_I,\ldots,k_r)$ and $(k'_I,\ldots,k'_r)$ are equal up to a permutation then the modules $\mathcal{J}\!\mathcal{L}_{k_I,\ldots,k_r}$ and $\mathcal{J}\!\mathcal{L}_{k'_I,\ldots,k'_r}$ are isomorphic.

I2.5. By means of the Frobenius reciprocity we can reformulate the Gelfand-Graev theorem (see 9.4) as follows: any non-degenerate irreducible representation of G_n occurs in $\mathcal{J}\!\mathcal{L}_n$ with multiplicity I. We extend this result to all irreducible representations of G_n.

<u>Proposition</u>. For any $\omega \in \Omega(G_n)$ there exists a degenerate Gelfand-Graev module $\mathcal{J}\!\mathcal{L}$, containing ω with multiplicity I. More precisely, let $\omega = \{\varphi\}$, where $\varphi \in S(\mathcal{C}\,;\,\mathcal{P})$, and $\mathrm{c.f.}(\varphi(\rho)^t) = (m_I(\rho), m_2(\rho),\ldots)$ for all $\rho \in \mathcal{C}$. Then we can choose $\mathcal{J}\!\mathcal{L} = \mathcal{J}\!\mathcal{L}_{k_I,\ldots,k_r}$, where

$$k_i = \sum_{\rho\in\mathcal{C}} m_i(\rho)\cdot\deg\rho \qquad (i = I,2,\ldots)$$

This follows at once from Theorem I2.I and the fact that

$$\langle\{\mu\}, y_\lambda\rangle = \begin{cases} 0 & \text{if c.f.}(\lambda) \text{ is lexicographically higher than c.f.}(\mu^t); \\ 1 & \text{if } \lambda = \mu^t \end{cases}$$

(see 4.I).

Q.E.D.

I2.6. By a degenerate Gelfand-Graev model of $\omega \in \Omega(G_n)$ we mean a realization of ω as a submodule of the module $\pi_{k_1,\dots,k_r}$ from I2.5. This is an analogue of the degenerate Whittaker model for representations of the groups GL(n) over a $\mathfrak{p}$-adic field, obtained in [2]. Let us give an application of this model.

Proposition. The Schur index of any irreducible representation of G_n equals I.

Proof. According to the theorem by R.Gow [28], any degenerate Gelfand-Graev module is rational i.e. is defined over $\mathbb{Q}$. We recall that the Schur index of ω is defined as $[K_1 : K_2]$ where K_1 is the field of definition of ω, and K_2 is the minimal field containing all character values of ω. It follows that if ω occurs with multiplicity I in a rational representation π then the Schur index of ω is I (indeed, the projection of π onto ω is just the character of ω, considered as an element of the group algebra; see e.g. [9]). Therefore, our assertion follows at once from I2.5.

Q.E.D.

When char $\mathbb{F}_q \neq 2$, the proposition is proved by Z.Ohmori [I6]. The proof in [I6] is also based on the R.Gow theorem but it is rather roundabout.

§ 13. Representations of general affine groups and the branching rule

13.1. Denote by P_n the subgroup in G_n consisting of matrices with the last row $(0,0,\dots,0,1)$; evidently, P_n can be realized as the group of all affine transformations of an $(n-1)$-dimensional affine space over $\mathbb{F}_q$. In this section we classify irreducible representations of P_n and describe the restriction of irreducible representations of G_n to P_n, and that of P_n to G_{n-1}.

Clearly, P_n has the abelian normal subgroup $V_n = U_{n-1,1}$ (see 9.1), and decomposes into the semidirect product $P_n = G_{n-1} \cdot V_n$. It is easy to verify that the set of characters of V_n has exactly two orbits under the action of G_{n-1}, namely the orbit of the identity character 1 and that of the character ψ (see 9.4). Clearly,

$$\operatorname{stab}_{G_{n-1}} 1 = G_{n-1}, \quad \operatorname{stab}_{G_{n-1}} \psi = P_{n-1} .$$

The next proposition follows at once from the well-known description of irreducible representations of a semidirect product $G \cdot V$, where V is an abelian normal subgroup in $G.V$ (see [9], 9.2).

<u>Proposition</u>. The operators

$$r_{V_n,1} \oplus r_{V_n,\psi} : R(P_n) \longrightarrow R(G_{n-1}) \oplus R(P_{n-1})$$

and

$$i_{V_n,1} \oplus i_{V_n,\psi} : R(G_{n-1}) \oplus R(P_{n-1}) \longrightarrow R(P_n)$$

(see 8.I) are mutually inverse isomorphisms between T-groups $R(P_n)$ and $R(G_{n-I}) \oplus R(P_{n-I})$.

I3.2. Applying Prop. I3.1 successively to $P_n, P_{n-I}, \dots P_I$, we obtain that $R(P_n)$ is isomorphic as a T-group to

$$R(G_{n-I}) \oplus R(G_{n-2}) \oplus \dots \oplus R(G_o).$$

More precisely, let us define the operators

$$R = R_n : R(P_n) \longrightarrow \bigoplus_{k=I}^{n} R(G_{n-k}) \quad \text{and}$$

$$I = I_n : \bigoplus_{k=I}^{n} R(G_{n-k}) \longrightarrow R(P_n)$$

by induction on n as follows:

R_n is the composition

$$R(P_n) \xrightarrow{r_{V_n,I} \oplus r_{V_n,\psi}} R(G_{n-I}) \oplus R(P_{n-I}) \xrightarrow{id \oplus R_{n-I}}$$

$$\longrightarrow R(G_{n-I}) \oplus \bigoplus_{k=1}^{n-1} R(G_{n-I-k}) = \bigoplus_{k=1}^{n} R(G_{n-k})$$

(the definition of I_n is quite similar).

<u>Proposition</u>. R and I are mutually inverse isomorphisms between T-groups $R(P_n)$ and $\bigoplus_{k=1}^{n} R(G_{n-k})$.

This follows at once from I3.I.

In particular, irreducible representations of P_n are in one-to-one correspondence with irreducible representations of the groups $G_{n-I}, G_{n-2}, \dots, G_o$. By 9.4, they are parametrized by the set

$$\coprod_{k<n} S_k(\mathcal{C}; \mathcal{P});$$

the representation of P_n, corresponding to a function φ, is $I_n\{\varphi\}$

13.3. The next result will play the crucial role in our description of restriction of representations of G_n to P_n and that of P_n to G_{n-1}.

<u>Proposition.</u> (a) The composition

$$R \circ \operatorname{Res}^{G_n}_{P_n} : R(G_n) \longrightarrow \bigoplus_{k=1}^{n} R(G_{n-k})$$

equals $D-I$ (where D is defined in 12.2).

(b) The composition

$$R \circ \operatorname{Ind}^{P_n}_{G_{n-1}} : R(G_{n-1}) \longrightarrow \bigoplus_{k=1} R(G_{n-k})$$

equals D.

<u>Proof</u>. (a) Let us compare the components of the operators $R \circ \operatorname{Res}^{G_n}_{P_n}$ and $D-I$, acting from $R(G_n)$ to $R(G_{n-k})$. By definition of R (see 13.2) the component of $R \circ \operatorname{Res}^{G_n}_{P_n}$ is the composition

$$r_{V_{n-k+1},1} \circ r_{V_{n-k+2},\psi} \circ r_{V_{n-k+3},\psi} \circ \dots \circ r_{V_n,\psi} .$$

On the other hand, the corresponding component of $D-I$, i.e. D_k, equals the composition

$$r_{U'_k,\psi} \circ r_{U_{n-k,k},1} \qquad \text{(see the proof of 12.2).}$$

Each of these compositions can be computed by means of 8.1 (c); we obtain that they coincide, as desired.

(b) First we compute the compositions

$$r_{V_n,1} \circ \mathrm{Ind}_{G_{n-1}}^{P_n} : R(G_{n-1}) \longrightarrow R(G_{n-1})$$

and

$$r_{V_n,\psi} \circ \mathrm{Ind}_{G_{n-1}}^{P_n} : R(G_{n-1}) \longrightarrow R(P_{n-1}).$$

Applying the general theorem on the composition $r \circ \mathrm{i}$ (see 8.3), we see that the first composition is the identity while the second one equals $\mathrm{Res}_{P_{n-1}}^{G_{n-1}}$; as usual the details will be given in Appendix 3 (see A3.7).

Remembering the definition of R, we see that the composition

$$R \circ \mathrm{Ind}_{G_{n-1}}^{P_n} : R(G_{n-1}) \longrightarrow R(G_{n-1}) \oplus \bigoplus_{k=1}^{n-1} R(G_{n-1-k})$$

equals $\mathrm{id} \oplus (R \circ \mathrm{Res}_{P_{n-1}}^{G_{n-1}})$. It remains to apply (a).

Q.E.D.

13.4. As a first application of Prop.13.3, we give another proof of the S.I.Gelfand theorem, i.e. of the inequality $\delta(\rho) \neq 0$ $(\rho \in \mathcal{C})$; see 9.4, 11.4. Since ρ is primitive, the definition of D (see 12.2) implies that

$$(D-1)\rho = \delta(\rho) \in R(G_0) = \mathbb{Z}$$

But by 13.3 (a),

$$(D-1)\pi \neq 0 \quad \text{for any non-zero} \quad \pi \in \mathcal{A}(G_n) .$$

Therefore, $\delta(\rho) \neq 0$, as desired.

13.5. The next theorem gives a complete description of restriction of irreducible representations of G_n to P_n and of P_n to G_{n-1}. For $\varphi, \varphi' \in S(\mathcal{C};\mathcal{P})$ we write $\varphi' \dashv \varphi$ if $\varphi'(\rho) \dashv \varphi(\rho)$ for all $\rho \in \mathcal{C}$ (see 4.3).

<u>Theorem</u>. (a) If $\varphi \in S_n(\mathcal{C};\mathcal{P})$ then the restriction of the irreducible representation $\{\varphi\}$ of G_n to P_n equals

$$\oplus \; I_n\{\varphi'\},$$

the sum is over all $\varphi' \dashv \varphi$ except $\varphi' = \varphi$.

(b) If $\varphi \in S(\mathcal{C};\mathcal{P})$ and $\deg \varphi < n$ then the restriction of the irreducible representation $I_n\{\varphi\}$ of P_n to G_{n-1} equals

$$\oplus \{\overline{\varphi}\},$$

the sum is over $\overline{\varphi} \in S_{n-1}(\mathcal{C};\mathcal{P})$ such that $\varphi \dashv \overline{\varphi}$.

In particular, the restriction of any irreducible representation of G_n (P_n) to P_n (G_{n-1}) is multiplicity - free.

<u>Proof</u>. By 12.3 and 4.3,

$$(*) \qquad D(\{\varphi\}) = \sum_{\varphi' \dashv \varphi} \{\varphi'\}.$$

With the account of (*), each assertion to be proved follows from the corresponding part of Prop. 13.3 (to prove (b) one has to apply the Frobenius reciprocity).

Q.E.D.

13.6. <u>Corollary</u>. (S.I.Gelfand [14]). The restriction of any cuspidal representation ρ of G_n to P_n is irreducible and equals

$$I_n\{\emptyset\} = \operatorname{Ind}_{U_n}^{P_n} \psi;$$

in particular, it does not depend on ρ .

I3.7. Corollary. An irreducible representation of G_n remains irreducible, being restricted to P_n, if and only if it has the form $\varphi_\rho(x_k)$, where $\rho \in \mathcal{C}$ and $n=k\cdot\deg\rho$.

This corollary shows another way to define the isomorphisms φ_ρ (see 9.5) i.e. another approach to the classification of irreducible representations of the groups G_n . Such an approach for the $\mathfrak{p}$ -adic groups was developed in [2] .

I3.8. Consider the chain of groups

$$G_0 \subset P_1 \subset G_1 \subset P_2 \subset G_2 \subset \ldots .$$

Theorem I3.5 says that the restriction of any irreducible representation of any group of the chain to the previous group has simple spectrum, and gives the explicit description of this spectrum. This is the branching rule , similar to that for the chain of the groups S_n (or, more generally, of the groups $S_n[G]$ where G is abelian; see 6.4 (2),7.9); cf. also the Gelfand-Cetlin Rule for the chain of unitary groups.

Restricting an irreducible representation of G_n to P_n, and then to G_{n-1}, we obtain.

Corollary. Let $\varphi \in S_n(\mathcal{C};\mathcal{P})$ and $\varphi' \in S_{n-1}(\mathcal{C};\mathcal{P})$. The multiplicity of $\{\varphi'\} \in \Omega(G_{n-1})$ in the restriction of $\{\varphi\} \in \Omega(G_n)$ to G_{n-1}, equals the number of $\varphi'' \in S(\mathcal{C};\mathcal{P})$ such that $\varphi'' \dashv \varphi$ and $\varphi'' \dashv \varphi'$.

This result in another form and by a quite different method was obtained by E.Thoma [I8] .

Appendix 1. Elements of the Hopf algebra theory

AI.I. For convenience we collect together the general results on Hopf algebras, needed in this paper. We use the notation and terminology from I.3-I.5.

Theorem. Let $\mathcal{A}$ be a connected quasi-Hopf algebra over a field K of characteristic 0, satisfying

(*) $I = P \oplus I^2$ (see I.5).

Then $\mathcal{A}$ is the symmetric algebra of the subspace P, i.e. $\mathcal{A}$ is a polynomial algebra in any basis of P. In particular, $\mathcal{A}$ is a Hopf algebra with commutative multiplication and comultiplication.

Theorem is proved in AI.3-AI.5.

AI.2. Remarks. (a) Suppose that any homogeneous component of $\mathcal{A}$ is finite-dimensional, and that $\mathcal{A}$ has a non-degenerate inner product satisfying the self-adjointness condition (S) from I.4. According to I.7, these assumptions imply (*), so Theorem AI.I is applicable. In this work we use Theorem AI.I only in such a situation.

(b) It is proved in § 4 of [8] that (*) and so the conclusion of Theorem AI.I is equivalent to the fact that $\mathcal{A}$ is a Hopf algebra with commutative multiplication and comultiplication.

AI.3. Lemma. Let $\mathcal{A}$ be a connected quasi-Hopf algebra over arbitrary commutative ring, such that

$$P \cap I^2 = 0 .$$

Then the multiplication in $\mathcal{A}$ is commutative and associative.

Proof. For $x, y \in \mathcal{A}$ put $[x,y] = xy - yx$.

Commutativity of the multiplication means that $[x,y] \equiv 0$. Since $\mathcal{A}$ is connected, the subspace $\mathcal{A}_0$ lies in the centre of $\mathcal{A}$, so we assume that $x \in \mathcal{A}_k$, $y \in \mathcal{A}_l$ where $k,l > 0$. By the Hopf axiom (H) we have

$$m^*[x,y] = [m^*x, m^*y] = [x \otimes I + I \otimes x + m_+^*(x),\ y \otimes I + I \otimes y + m_+^*(y)].$$

Using induction on k+l, we can assume that $m_+^*(x)$ commutes with each of the elements $y \otimes I$, $I \otimes y$, and $m_+^*(y)$, while $m_+^*(y)$ commutes with $x \otimes I$ and $I \otimes x$. Hence,

$$m^*[x,y] = [x \otimes I + I \otimes x,\ y \otimes I + I \otimes y] = [x,y] \otimes I + I \otimes [x,y],$$

so $[x,y] \in P$. On the other hand, $[x,y] \in I^2$, hence $[x,y] = 0$, as desired. The proof of associativity is similar (one has only to consider $x(yz) - (xy)z$ instead of $[x,y]$).

Q.E.D.

AI.4. Now let $\mathcal{A}$ satisfy the hypothesis of Theorem AI.I. Denote by S(P) the symmetric algebra of P. Consider the comultiplication on S(P) satisfying (H) and such that all elements of P are primitive. Clearly, S(P) becomes a Hopf algebra with commutative multiplication and comultiplication. By AI.3, the multiplication in $\mathcal{A}$ is commutative and associative, so we obtain the natural Hopf algebra morphism

$$p : S(P) \longrightarrow \mathcal{A}$$

We must prove that p is an isomorphism.

The fact that p is an epimorphism, follows in a standard way from the fact that P is mapped epimorphically onto I/I^2. To prove that p is a monomorphism we use the next.

Lemma. The subspace of primitive elements of the Hopf algebra S(P) coincides with P.

Let us derive from this Lemma that p is monomorphic. Suppose that $\operatorname{Ker} p \neq 0$ and let u be a non-zero homogeneous element in $\operatorname{Ker} p$ of least possible degree. Clearly, $\deg u > 0$ and $u \notin P$. By Lemma, $m_+^*(u) \neq 0$. Since u is an element with the least degree in $\operatorname{Ker} p$, it follows that

$$(p \otimes p)\,(m_+^*(u)) \neq 0.$$

Since p is a Hopf algebra morphism, we have

$$(p \otimes p)\,(m_+^*(u)) = m_+^*\,(p(u)),$$

hence $p(u) \neq 0$. We obtain a contradiction.

AI.5. <u>Proof of Lemma AI.4</u>. Denote by $S^n(P)$ the n-th symmetric power of P, i.e. the space spanned by the products $v_I v_2 \ldots \cdot v_n$, $v_i \in P$. For any $u \in S^n(P)$ denote by $m_I^*(u)$ the component of $m^*(u)$, belonging to $S^{n-I}(P) \otimes S^I(P)$. It suffices to verify that

$$m_I^* : S^n(P) \longrightarrow S^{n-I}(P) \otimes S^I(P)$$

is a monomorphism for $n \geqslant 2$. This in turn follows at once from the identity

$$(*) \qquad m \circ m_I^*\,(u) = nu \qquad \text{for } u \in S^n(P),$$

where m stands for multiplication (note that here the assumption char $K = 0$ is used).

It suffices to verify $(*)$ for $u = v_I \ldots v_n$, where $v_i \in P$. We have

$$m^*(u) = \prod_{i=I}^{n} m^*(v_i) = \prod_{i=I}^{n} (v_i \otimes I + I \otimes v_i),$$

hence

$$m_I^*(u) = \sum_{i=1}^{n} v_I v_2 \dots v_{i-I}\, v_{i+I} \dots v_n \otimes v_i \quad ,$$

and (*) follows.

Q.E.D.

Theorem AI.I is done.

AI.6. Now we consider the conjugation of a Hopf algebra.

Proposition. Let $\mathcal{A}$ be a connected Hopf algebra over a commutative ring with unit K.

(a) There exists the unique morphism of graded K-modules $T = T_{\mathcal{A}} : \mathcal{A} \longrightarrow \mathcal{A}$ such that the diagram

$$\begin{array}{ccccccc} \mathcal{A} & \xrightarrow{m^*} & \mathcal{A}\otimes\mathcal{A} & \xrightarrow{id\otimes T} & \mathcal{A}\otimes\mathcal{A} & \xrightarrow{m} & \mathcal{A} \\ & {\scriptstyle e^*}\searrow & & K & & \nearrow{\scriptstyle e} & \end{array}$$

is commutative. T is called the conjugation of $\mathcal{A}$.

(b) If $\mathcal{A}$ has commutative multiplication and comultiplication then T is an involutive Hopf algebra automorphism of $\mathcal{A}$.

This Proposition is proved in AI.8-AI.IO.

AI.7. Remark. In fact, part (b) can be generalized as follows. For any connected Hopf algebra $\mathcal{A}$ the conjugation T is an antiautomorphism of $\mathcal{A}$; if either multiplication or comultiplication in $\mathcal{A}$ is commutative then T is involutive i.e. $T^2 = id$ (see [8], 8.6-8.8). We shall follow lines of [8], § 8 with some simplifications caused by the commutativity of multiplication and comultiplication in $\mathcal{A}$.

AI.8. For any two connected Hopf algebras $\mathcal{A}$ and $\mathcal{B}$ over K denote by $G(\mathcal{A},\mathcal{B})$ the set of graded K-module morphisms

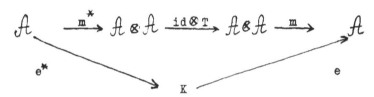

$f : \mathcal{A} \longrightarrow \mathcal{B}$, such that $f = \mathrm{id}$ on $\mathcal{A}_0 = K = \mathcal{B}_0$. For any $f, g \in G(\mathcal{A},\mathcal{B})$ denote by $f * g$ the composition:

$$\mathcal{A} \xrightarrow{m^*} \mathcal{A}\otimes\mathcal{A} \xrightarrow{f\otimes g} \mathcal{B}\otimes\mathcal{B} \xrightarrow{m} \mathcal{B}$$

<u>Proposition</u>. (a) $G(\mathcal{A},\mathcal{B})$ is a group under the operation $*$ with identity

$$\mathcal{A} \xrightarrow{e^*} K \xrightarrow{e} \mathcal{B}$$

(the morphism which is 0 on $\bigoplus_{n>0} \mathcal{A}_n$ and id on $\mathcal{A}_0$).

(b) If $f : \mathcal{A}' \longrightarrow \mathcal{A}$ is a coalgebra morphism and $g : \mathcal{B} \longrightarrow \mathcal{B}'$ is an algebra morphism then the mapping $G(f,g) : G(\mathcal{A},\mathcal{B}) \longrightarrow$ $\longrightarrow G(\mathcal{A}',\mathcal{B}')$, sending h to $g \circ h \circ f$, is a group homomorphism.

<u>Proof</u>. (a) Associativity of "$*$" can be verified directly (one must use that comultiplication in $\mathcal{A}$ and multiplication in $\mathcal{B}$ are associative). The fact that $e \circ e^*$ is the identity of $G(\mathcal{A},\mathcal{B})$, is obvious. It remains to prove the existence of the inverse element f^{-1} to any $f \in G(\mathcal{A},\mathcal{B})$. Let $n > 0$ and assume f^{-1} to be defined on $\bigoplus_{k<n} \mathcal{A}_k$. We define f^{-1} on $\mathcal{A}_n$ by

$$f^{-1}(u) = -f(u) - m\circ(f \otimes f^{-1}) \circ m_+^*(u) \qquad \text{for } u \in \mathcal{A}_n .$$

Clearly, $f^{-1} \in G(\mathcal{A},\mathcal{B})$ is well-defined and is inverse to f.

Part (b) follows at once from definitions.

Q.E.D.

AI.9. In terms of AI.8 the conjugation of a Hopf algebra $\mathcal{A}$ is the element of $G(\mathcal{A},\mathcal{A})$, inverse to the identity morphism $\mathrm{id} : \mathcal{A} \longrightarrow \mathcal{A}$. Therefore, AI.6 (a) follows from AI.8 (a). The next proposition follows at once from definitions.

<u>Proposition</u>. If $\mathcal{A}$ and $\mathcal{B}$ are connected Hopf algebras then

$$T_{\mathcal{A}} \otimes T_{\mathcal{B}} : \mathcal{A} \otimes \mathcal{B} \longrightarrow \mathcal{A} \otimes \mathcal{B}$$

is the conjugation of $\mathcal{A} \otimes \mathcal{B}$

AI.10. <u>Proof of AI.6 (b).</u> First we prove that if the comultiplication in $\mathcal{A}$ is commutative then

$$T : \mathcal{A} \longrightarrow \mathcal{A}$$

is a coalgebra morphism, i.e. the diagram

$$\begin{array}{ccc} \mathcal{A} & \xrightarrow{m^*} & \mathcal{A} \otimes \mathcal{A} \\ {\scriptstyle T}\downarrow & & \downarrow{\scriptstyle T \otimes T} \\ \mathcal{A} & \xrightarrow{m^*} & \mathcal{A} \otimes \mathcal{A} \end{array}$$

is commutative. Since m^* is an algebra morphism, by AI.8 (b) we have the group morphism

$$G(id, m^*) : G(\mathcal{A}, \mathcal{A}) \longrightarrow G(\mathcal{A}, \mathcal{A} \otimes \mathcal{A}).$$

It follows that the elements m^* and $m^* \circ T$ of $G(\mathcal{A}, \mathcal{A} \otimes \mathcal{A})$ are inverse to each other. It remains to verify that $(T \otimes T) \circ m^*$ also is inverse to m^*, i.e. that $m^* * \left[(T \otimes T) \circ m^*\right]$ is the identity element of $G(\mathcal{A}, \mathcal{A} \otimes \mathcal{A})$. By definition, $m^* * \left[(T \otimes T) \circ m^*\right]$ is the composition

$$(*) \begin{cases} \mathcal{A} \xrightarrow{m^*} \mathcal{A} \otimes \mathcal{A} \xrightarrow{m^* \otimes m^*} \mathcal{A} \otimes \mathcal{A} \otimes \mathcal{A} \otimes \mathcal{A} \xrightarrow{id \otimes T \otimes T} \mathcal{A} \otimes \mathcal{A} \otimes \mathcal{A} \otimes \mathcal{A} \\ \xrightarrow{\overline{m}} \mathcal{A} \otimes \mathcal{A}, \end{cases}$$

where $\overline{m}$ is the multiplication in $\mathcal{A} \otimes \mathcal{A}$. Commutativity of the comultiplication in $\mathcal{A}$ implies that

$$(m^* \otimes m^*) \circ m^* = \bar{m}^* \circ m^* ,$$

where $\bar{m}^*$ is the comultiplication in $\mathcal{A} \otimes \mathcal{A}$. Hence, $m^* * [(T \otimes T) \circ m^*] = [id_{\mathcal{A} \otimes \mathcal{A}} * (T \otimes T)] \circ m^*$.

By AI.8 (b), $id_{\mathcal{A} \otimes \mathcal{A}} * (T \otimes T)$ is the identity element of $G(\mathcal{A} \otimes \mathcal{A}, \mathcal{A} \otimes \mathcal{A})$ so its composition with m^* is the identity element of $G(\mathcal{A}, \mathcal{A} \otimes \mathcal{A})$, as desired.

One can prove in a quite similar way, that if the multiplication in $\mathcal{A}$ is commutative then T is an algebra morphism. It remains to prove that $T^2 = id$. It suffices to verify that $T^2 \in G(\mathcal{A}, \mathcal{A})$ is inverse to T, i.e. that

$$T * T^2 = e \circ e^* .$$

By definition,

$$T * T^2 = m \circ (T \otimes T) \circ (id_{\mathcal{A}} \otimes T) \circ m^* .$$

Since T is an algebra morphism, we have

$$m \circ (T \otimes T) = T \circ m ,$$

hence

$$T * T^2 = T \circ [id_{\mathcal{A}} * T] = T \circ e \circ e^* = e \circ e^* .$$

Q.E.D.

Appendix 2. A combinatorial proposition.

A2.I. In this Appendix we complete the proof of Prop. 4.I8. Let us give another combinatorial definition of the coefficient $g_{\varkappa,\nu}$ (see 4.I8). Define the linear order relation "$\leq_J$" on $\mathbb{N} \times \mathbb{N}$ as follows:

$(i,j) <_J (i',j')$ if and only if either $i < i'$, or $i = i'$, and $j > j'$.

Let $\mathcal{X}$ be a subset of $\mathbb{N}\times\mathbb{N}$ and f a mapping from $\mathcal{X}$ to $\mathbb{N}\times\mathbb{N}$. We say that f satisfies (J) if f is a morphism of ordered sets $(\mathcal{X}, \leq_P) \longrightarrow (\mathbb{N}\times\mathbb{N}, \leq_J)$ i.e.

if $x, x' \in \mathcal{X}$ and $x \leq_P x'$ then $f(x) \leq_J f(x')$,

(see 4.I2). We call <u>a picture</u> any bijection $f : \mathcal{X}_1 \xrightarrow{\sim} \mathcal{X}_2$ between two skew diagrams such that f and the inverse bijection f^{-1} both satisfy (J); denote by $\mathcal{P}(\mathcal{X}_1, \mathcal{X}_2)$ the set of pictures $f : \mathcal{X}_1 \xrightarrow{\sim} \mathcal{X}_2$.

It turns out that $g_{\mathcal{X},\nu}$ equals the number of pictures $f : \nu \xrightarrow{\sim} \mathcal{X}$. The proof of this fact will be sketched in A2.3(b). Note that this is not significant for us: we include the present formulation of 4.I8 only since it is classical. Now, in principle, the reader may forget it; we shall prove that the inner product $\langle\{\nu\}, \{\mathcal{X}\}\rangle$ equals the number of pictures $f : \nu \xrightarrow{\sim} \mathcal{X}$ Remembering 4.I8 (**), we see that this follows at once from the next combinatorial.

<u>Proposition</u>. Let ν be a Young diagram and $\mathcal{X}$ a skew diagram such that $|\nu| < |\mathcal{X}|$. There exists a bijection between the sets

$$\coprod_{\mathcal{X}' \dashv \mathcal{X}} \mathcal{P}(\nu, \mathcal{X}') \quad \text{and} \quad \coprod_{\nu \dashv \bar{\nu}} \mathcal{P}(\bar{\nu}, \mathcal{X}).$$

The remainder of this Appendix is devoted to the proof of this proposition.

A2.2. First, we reformulate the definition of a picture in a more "working" manner. Any point $a \in \mathbb{N}\times\mathbb{N}$ divides $\mathbb{N}\times\mathbb{N} \setminus \{a\}$ into 8 regions shown at the figure:

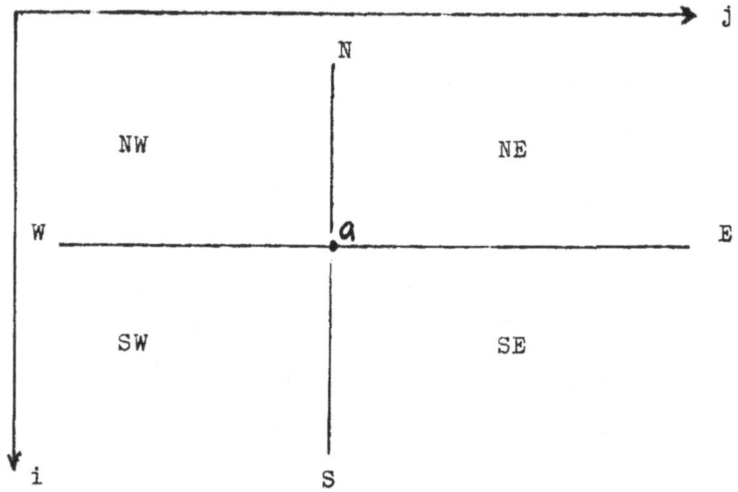

We shall write $b(X,Y,\dots)a$ if $b \neq a$ and b lies in one of the regions $X,Y,\dots$ with respect to a.

For example, the relation $b_J > a$ ($b_P > a$) means that

$$b(W,SW,S,SE)a \qquad (b(S,SE,E)a).$$

Lemma. Let $f : \mathcal{X}_1 \to \mathcal{X}_2$ be a bijection between two skew diagrams $\mathcal{X}_1$ and $\mathcal{X}_2$. Then f is a picture if and only if it satisfies the following implications ($x,x' \in \mathcal{X}_1$) :

(I) If $x(E)x'$ then $f(x)\,(W,SW)f(x')$

(2) If $x(S)x'$ then $f(x)\,(SW,S)f(x')$

(3) If $x(NE)x'$ then $f(x)\,(NE,N,NW,W,SW)f(x')$

(4) If $x(SE)x'$ then $f(x)\,(SW)f(x')$.

Proof. Clearly, f satisfies (J) if and only if

(J) If $x(S,SE,E)x'$ then $f(x)(W,SW,S,SE)f(x')$

Using the logical law

$$[A \Longrightarrow B] \Longleftrightarrow [\neg B \Rightarrow \neg A]$$

we see that f^{-1} satisfies (J) if and only if

(J^{-I}) If $x(E,NE,N,NW)x'$ then $f(x)\ (NE,N,NW,W,SW)f(x')$

By formal manipulations we see that f satisfies (J) and (J^{-I}) if and only if f satisties (I), (3), (4) and

(2') If $x(S)x$ then $f(x)\ (SW,S,SE)f(x')$.

It remains to verify that (I),(2'), (3) and (4) imply (2). Suppose, (2) does not hold. Then there exist two points $x'=(i,j)$ and $x = (i+I,j)$ from $\mathcal{X}_1$ aush that $f(x)\ (SE)f(x')$. Consider the point y such that

$$y(S)f(x') \quad \text{and} \quad f(x)(E)\ y \quad .$$

Since $\mathcal{X}_2$ is a skew diagram, $y \in \mathcal{X}_2$, i.e. $y = f(x")$ for some $x" \in \mathcal{X}_1$. By (J^{-I}):

$$x' <_J x" <_J x \quad .$$

It follows that either $x'(E)x"$ or $x(W)x"$.

<u>Case I:</u> $x'(E)x"$. Consider the point x^0 such that $x^0(S)x"$ and $x(E)x^0$. Again, we see that $x^0 \in \mathcal{X}_1$ since $\mathcal{X}_1$ is a skew diagram. By (I) and (2'),

$$f(x^0)\ (E,NE)f(x) \quad \text{and} \quad f(x^0)\ (SW,S,SE)f(x").$$

This contradicts the fact that $f(x)$ and $f(x") = y$ lie in the same row.

<u>Case 2:</u> $x(W)x"$. Again consider the fourth vertex x^0 of the rectangular $\{x',x,x",x^0\}$. Applying (I) ((2')) to x^0 and x' (x^0 and $x"$), we see that

$$f(x^0)\ (NE,N,NW)f(x") \quad \text{and} \quad f(x^0)\ (W,SW)f(x').$$

Since $f(x')$ and $f(x")$ lie in the same column, it follows that

$$f(x^0)\ (NW) f(x'').$$

Thus, the relative position of x^0 and x'' is the same as that of x' and x, i.e.

$$x''(S)x^0 \quad \text{and} \quad f(x'')\ (SE) f(x^0).$$

But the pair $\{x'',x^0\}$ lies to the east of $\{x',x\}$. Since $\mathcal{X}_1$ is finite, this again leads to a contradiction.

Q.E.D.

A2.3. Remarks. (a) The author has learnt the notion of a picture from [19], where it was defined as a bijection between two finite subsets of $\mathbb{N}\times\mathbb{N}$, satisfying (1) through (4). Our definition is more symmetric; it implies at once the remarkable duality, namely that the bijection inverse to a picture is itself a picture. Note that in our situation (when $\mathcal{X}_1$ and $\mathcal{X}_2$ are skew diagrams) (4) follows from (1) and (2).

(b) Now we sketch the proof of the fact that $g_{\mathcal{X},\mathcal{Y}}$ from 4.18 equals the number of pictures $f:\mathcal{Y} \rightleftarrows \mathcal{X}$. Assign to each numbering $\varphi:\mathcal{X}\to\mathbb{N}$ of type c.f.($\mathcal{Y}$) (see 4.14) the bijection $f:\mathcal{Y}\rightleftarrows\mathcal{X}$, uniquely determined by the properties:

$$\varphi(f(i,j)) = i \quad \text{and} \quad f(i,j) <_J f(i,j+1) \quad (\text{for } (i,j),\ (i,j+1) \in \mathcal{Y}).$$

It is easy to see that the correspondence $\varphi \mapsto f$ is well-defined and injective, and that the condition (J) from 4.18 means just that f satisfies (J). As in A2.2, one can express this condition as well as the condition that φ is a column-strict numbering, in terms of implications similar to (1)-(4). The formal manipulations show that these conditions are equivalent to the fact that f satisfies (1), (2'), (3) and (4).

It remains to apply A2.2.

A2.4. We say that a point b of a skew diagram $\mathcal{X}$ is <u>regular</u> if b is a maximal element of $\mathcal{X}$ relative to "$\leq_P$" (cf. 4.12). For any $x = (i,j) \in \mathbb{N} \times \mathbb{N}$ we write $i = pr_1 x$ and $j = pr_2 x$.

Our proof of Prop. A2,1 is based on the following two algorithms.

<u>Algorithm E.</u> (Extension of a picture). Data: a skew diagram $\mathcal{X}$, a regular point $b \in \mathcal{X}$, a Young diagram $\mathcal{Y}$, and a picture $f: \mathcal{Y} \rightleftarrows \mathcal{X} \setminus \{b\}$. The algorithm constructs the Young diagram $\overline{\mathcal{Y}}$ obtained by adding one point to $\mathcal{Y}$, and the picture $\bar{f}: \overline{\mathcal{Y}} \rightleftarrows \mathcal{X}$.

<u>Algorithm R</u> (Reduction of a picture). Data: a skew diagram $\mathcal{X}$, a Young diagram $\overline{\mathcal{Y}}$ with a regular point $a \in \overline{\mathcal{Y}}$, and a picture $\bar{f}: \overline{\mathcal{Y}} \rightleftarrows \mathcal{X}$. The algorithm constructs the skew diagram $\mathcal{X}'$ obtained by taking away the regular point b of $\mathcal{X}$ and the picture $f: \overline{\mathcal{Y}} \setminus \{a\} \rightleftarrows \mathcal{X}'$.

<u>Description of Algorithm E</u>. For any $j \in \mathbb{N}$ let C_j be the j-th column of $\mathcal{Y}$ and n_j the length of C_j. By (J):

$$f(1,j) <_J f(2,j) <_J \dots <_J f(n_j,j) .$$

Put $b_0 = b$ and define successively the points $a_1, b_1, a_2, b_2, \dots, a_l, b_l$ as follows. If $j \geq 1$ then a_j is the point in C_j with the minimal value of pr_1 such that

$$f(a_j) >_J b_{j-1} ,$$

while $b_j = f(a_j)$. The process stops when

$$b_l >_J f(x) \quad \text{for all} \quad x \in C_{l+1} .$$

Put

$$a = a_{l+1} = (n_{l+1} + 1,\ l+1)\ ,\quad \bar{\mathcal{V}} = \mathcal{V} \cup \{a\},$$

and define the mapping $\bar{f}: \bar{\mathcal{V}} \longrightarrow \mathscr{X}$ by

$$\bar{f}(x) = \begin{cases} f(x) & \text{if } x \in \mathcal{V} \setminus \{a_1, \ldots, a_l\}\ ; \\ b_{j-1} & \text{if } x = a_j \quad (j = 1,2,\ldots,l+1) \end{cases}$$

Evidently, $\bar{f}$ is a bijection of $\bar{\mathcal{V}}$ onto $\mathscr{X}$. The facts that $\bar{\mathcal{V}}$ is a Young diagram and $\bar{f}$ is a picture, will be verified in the next item.

<u>Description of Algorithm R</u>. Let $pr_2\, a = l+1$ $(l \geq 0)$. Put $a_{l+1} = a$ and define successively the points $b_l, a_l, b_{l-1}, a_{l-1}, \ldots, b_1, a_1, b_0$, where $b_j \in \mathscr{X}$ and $a_j \in \bar{\mathcal{V}}$, as follows:

$$b_j = \bar{f}(a_{j+1}) \qquad (0 \leq j \leq l),$$

while a_j for $1 \leq j \leq l$ is the element of C_j with the maximal value of pr_1 such that $\bar{f}(a_j) <_J b_j$. It is easy to see that a_j is well defined i.e. for $1 \leq j \leq l$ there exists $x_j' \in C_j$ such that $\bar{f}(x_j') <_J b_j$ (one can take $x_j' = (pr_1\, a_{j+1}, j)$). Put

$$b = b_0,\quad \mathscr{X}' = \mathscr{X} \setminus \{b\},$$

and define the mapping $f: \bar{\mathcal{V}} \setminus \{a\} \longrightarrow \mathscr{X}'$ by

$$f(x) = \begin{cases} \bar{f}(x) & \text{if } x \in \bar{\mathcal{V}} \setminus \{a_1, \ldots, a_{l+1}\}\ ; \\ b_j & \text{if } x = a_j \quad (j = 1,2,\ldots,l) \end{cases}$$

Evidently, f is a bijection of $\overline{\mathcal{V}} \setminus \{a\}$ onto $\mathcal{X}'$. We leave to the reader to verify that b_0 is a regular point of $\mathcal{X}$, and f is a picture; this can be done in a quite similar way as for Algorithm E (cf. the next item).

By definitions, Algorithms E anf R are inverse to each other in the obvious sense. In particular, this proves Prop. A2,I in the case when $|\mathcal{X}| = |\mathcal{V}| + I$. In the general case, one has to use the iteration of these algorithms. More precisely, let $f:\mathcal{V} \rightrightarrows \mathcal{X}'$ be a picture, $\mathcal{X}' \dashv \mathcal{X}$, and $\mathcal{X} \setminus \mathcal{X}' = \{b^{(I)}, b^{(2)}, \ldots, b^{(k)}\}$, where

$$pr_I b^{(I)} < pr_I b^{(2)} \quad \ldots < pr_I b^{(k)} \quad .$$

Put $\mathcal{X}_i = \mathcal{X}' \cup \{b^{(I)}, b^{(2)}, \ldots, b^{(i)}\}$ $(0 \leq i \leq k)$. It is easy to see that each $\mathcal{X}_i$ is a skew diagram and $b^{(i)}$ is a regular point of $\mathcal{X}_i$. Let us define the pictures $f_i:\mathcal{V}_i \rightrightarrows \mathcal{X}_i$ (i= = 0,...,k) by induction on i as follows. We set $f_0 = f$, and define f_i for $i \geqslant I$ to be the picture obtained by applying Algorithm E to f_{i-I} and $b^{(i)}$. Let $\mathcal{V}_i \setminus \mathcal{V}_{i-1} = \{a^{(i)}\}$ $(I \leqslant i \leqslant k)$; one can easily verify that

$$pr_I a^{(I)} < pr_I a^{(2)} < \ldots < pr_I a^{(k)} \quad .$$

Set $\overline{\mathcal{V}} = \mathcal{V}_k$ and $\overline{f} = f_k$, i.e. $\overline{f} \in \mathcal{P}(\overline{\mathcal{V}}, \mathcal{X})$. Clearly, $\mathcal{V} \dashv \overline{\mathcal{V}}$; the correspondence $f \longmapsto \overline{f}$ is just that of Prop. A2,I. The inverse mapping $\overline{f} \longrightarrow f$ can be constructed in a quite similar way, by means of Algorithm R. This proves Prop. A2,I and hence Prop. 4.I8.

A2.5. It remains to verify that Algorithm E is well-defined, i.e. that $\overline{\mathcal{V}}$ constructed in its description is a Young diagram, while $\overline{f}$ is a picture. We proceed in a series of steps.

(I) Let us prove that

$$pr_I a_I \geqslant pr_I a_2 \geqslant \dots \geqslant pr_I a_{l+I} \quad ;$$

in particular, this implies that $\overline{\mathcal{V}}$ is a Young diagram and a_{l+I} is a regular point of $\overline{\mathcal{V}}$.

Suppose $pr_I a_{j-I} < pr_I a_j$ for some j. Consider $x = (pr_I a_{j-I}, j) \in C_j$. Since $a_{j-I} <_P x$, the condition (J) implies that $f(x) \; {}_J{>} f(a_{j-I}) = b_{j-I}$. This contradicts the choise of a_j.

(2) Let us prove that $\overline{f}$ satisfies (J), i.e. that if $x', x'' \in \overline{\mathcal{V}}$ and $x' >_P x''$ then $\overline{f}(x') \; {}_J{>} \overline{f}(x'')$. We can assume that x' and x'' are neighbours with respect to „$\leq_P$", i.e.

either $x' = (pr_I x'' + I, pr_2 x'')$ or $x' = (pr_I x'', pr_2 x'' + I)$.

If x' and x'' both don't lie in $\{a_I, \dots, a_{l+I}\}$ then

$$\overline{f}(x') = f(x') \; {}_J{>} f(x'') = \overline{f}(x'').$$

If x' and x'' both lie in $\{a_I, \dots, a_{l+I}\}$ then $x' = a_{j+I}$ and $x'' = a_j$ for some j, therefore

$$\overline{f}(x') = b_j \; {}_J{>} b_{j-I} = \overline{f}(x'') \quad .$$

Finally, let exactly one of x' and x'' lie in $\{a_I, \dots, a_{l+I}\}$. Since $\overline{f} \leq_J f$ on $\{a_I, \dots, a_{l+I}\}$, the case when $x'' \in \{a_I, \dots, a_{l+I}\}$ is obvious. It remains to prove that if $a_j = (i,j)$ and $a_{j-I} \neq (i, j-I)$ then

$$b_{j-I} \; {}_J{>} f(i-I, j) \quad \text{and} \quad b_{j-I} \; {}_J{>} f(i, j-I) \quad .$$

The first inequality follows at once from the definition of a_j; furthermore, by (I), $a_{j-I} \; {}_P{>} (i, j-I)$ hence

$$b_{j-1} = f(a_{j-1}) \ {}_J{>} f(i,j-1) ,$$

as desired.

(3) It remains to prove that $\bar{f}^{-1} : \mathcal{X} \rightarrowtail \bar{V}$ satisfies (J). As in the previous step, it suffices to prove that $\bar{f}^{-1}(y') \ {}_J{>} \bar{f}^{-1}(y'')$ whenever $y' \ {}_P{>} y''$, and y' and y'' are neighbours with respect to "$\leq_P$". The case when y' and y'' both don't lie in $\{b_0,\ldots,b_l\}$, is obvious. Let us treat the case when y' and y'' both lie in $\{b_0,\ldots,b_l\}$. We have

$$\bar{f}^{-1}(b_j) = a_{j+1} .$$

Since

$$a_1 >_J a_2 >_J \ldots >_J a_{l+1} \quad \text{(see (1))},$$

we must prove that the inequality $b_j <_P b_{j'}$ cannot hold for $0 \leq j < j' \leq l$. For $j = 0$ this follows from the fact that b_0 is a regular point of $\mathcal{X}$, while for $j > 0$ from the fact that f^{-1} satisfies (J) since

$$f^{-1}(b_j) = a_j \ {}_J{>} a_{j'} = f^{-1}(b_{j'}) \ .$$

Combining this assertion on the b_j's with the inequalities

$$b_0 <_J b_1 <_J \ldots <_J b_l ,$$

we obtain that

$$b_j (W,SW) \, b_{j-1} \quad \text{for } j = 1,\ldots,l .$$

(4) It remains to treat the case when exactly one of y' and y'' lies in $\{b_0,\ldots,b_l\}$. Since $\bar{f}^{-1} <_J f^{-1}$ on $\{b_0,\ldots,b_l\}$, the case when $y'' \in \{b_0,\ldots,b_l\}$, is obvious. Thus the only case to be considered is the following:

$y' = b_j$, $y'' \notin \{b_0,\dots,b_1\}$, and

y'' is either $(pr_I b_j - I,\ pr_2 b_j)$ or $(pr_I b_j, pr_2 b_j - I)$;

we must prove that $f^{-I}(y'') \lessdot_J a_{j+I}$. Put $x'' = f^{-I}(y'')$.

Let us prove that the inequality $x'' >_P a_{j+I}$ cannot hold. Indeed, this is clear for $j = 1$, since a_{1+I} is a regular point of $\overline{\mathcal{V}}$; but if $j < 1$ then $x_{j+I} \in \mathcal{V}$, hence by (J) the inequality $x'' >_P a_{j+I}$ would imply that

$$y'' = f(x'') >_J f(a_{j+1}) = b_{j+I} \quad .$$

This contradicts the condition $b_{j+I}(W,SW)b_j$ (see (3)).

(5) We have proved that

$$x''\ (NE,N,NW,W,SW)\ a_{j+I} \quad .$$

It remains to verify that the relation $x''(W,SW)\ a_{j+I}$ cannot hold. Assume the contrary, i.e. that

$$x''\ (W,SW)\ a_{j+I} \quad .$$

First, we see that $pr_2 x'' \leq j$; in particular, $j > 0$. Since $y'' <_P b_j$ and f^{-I} satisfies (J), we see that $x'' \lessdot_J a_j$. This implies $x''(N,NW)\ a_j$ hence $x'' \lessdot_P a_j$. Since f satisfies (J), it follows that $y'' <_J b_j$. Therefore, only one possibility for the relative position of y'' and b_j can hold (see (4)), namely

$$y'' = (pr_I b_j - I,\ pr_2 b_j\ ) \quad .$$

Denote this point y'' by b_j^-, and $f^{-I}(y'') = x''$ by x_j .

(6) This is just the time to apply Lemma A2.2 ! Applying (2) of A2.2 to the picture f^{-I} and the pair $b_j^-(N)b_j$, we obtain that

$$x_j (NE,N)\ a_j \ .$$

Comparing this with the relation $x_j(N,NW)\ a_j$ obtained in (5), we conclude that

$$x_j (N)\ a_j \ .$$

Thus, the desired contradiction follows from the next assertion

(*) For $j > 0$ the relation $x_j(N)\ a_j$ cannot hold.

Suppose $x_j(N)\ a_j$. By definition of a_j, we have

$$b_j^- = f(x_j) <_J b_{j-I} \ .$$

Combining this with the relation $b_{j-I}(E,NE)\ b_j$ proved in (3) we see that

$$b_{j-I}(E)\ b_j \ .$$

Now consider the point b_{j-I}^- (it lies in $\mathcal{X} \setminus \{b\}$, since $b_j^- \underset{P}{<} b_{j-I}^- \underset{P}{<} b_{j-I}$ and $\mathcal{X}$ is a skew diagram). Applying (I) of A2.2 to the picture f^{-I} and the pair $b_{j-I}^-(E)\ b_j^-$, we see that

$$x_{j-I}(W,SW)\ x_j \ .$$

In particular, $pr_2 x_{j-I} < pr_2 x_j = j$, so $j > I$. Therefore, $b_{j-I} \in \mathcal{X} \setminus \{b\}$.

Applying (2) of A2.2 to the picture f^{-I} and the pair $b_{j-I}^-(N)\ b_{j-I}$, we obtain

$$x_{j-I}(NE,N)\ a_{j-I} \ .$$

But $pr_2 x_{j-1} \leq j-I = pr_2 a_{j-I}$ hence

$$x_{j-I}(N)\ a_{j-I} \ .$$

Thus, under the assumption that $x_j(N)a_j$ we have proved that $j>I$ and $x_{j-I}(N)\ a_{j-I}$. The assertion (*) follows by induction on j.

Q.E.D.

A2.6. Remark. Our algorithms E and R are closely connected with the algorithms of insertion and deletion of a number into a Young tableau, playing the crucial role in the proof of the well-known Robinson-Schensted correspondence (see [30], 5.I.4). Our algorithms allow us to obtain a combinatorial generalization of the Robinson-Schensted correspondence and derive from it the following generalization of Prop. 4.I8:

for any two skew diagrams $\mathcal{X}_1$ and $\mathcal{X}_2$ the inner product $\langle\{\mathcal{X}_1\},\{\mathcal{X}_2\}\rangle$ equals the number of pictures $f:\mathcal{X}_1 \xrightarrow{\sim} \mathcal{X}_2$.

These questions are treated in detail in [20].

Appendix 3. The composition of functors r and i.

A3.I. Let G be a finite group, M, U, N, and V be subgroups in G, θ a character of U, and ψ a character of V. Suppose that $M\cap U = N\cap V = \{e\}$, M normalizes U and θ, while N normalizes V and ψ, i.e. there are defined the functors

$$i_{U,\theta} : \mathcal{A}(M) \longrightarrow \mathcal{A}(G) \quad \text{and}$$

$$r_{V,\psi} : \mathcal{A}(G) \longrightarrow \mathcal{A}(N) \qquad \text{(see 8.I)}.$$

Under the extra assumption we shall compute the composition

$$F = r_{V,\psi} \circ i_{U,\theta} : \mathcal{A}(M) \longrightarrow \mathcal{A}(N).$$

Put $P = MU$, $Q = NV$ and choose a set W of representatives of double cosets

$$Q \setminus G/P .$$

For any $w \in W$ denote also by w the corresponding inner automorphism of G, i.e. $w(g) = wgw^{-1}$; write $w(\theta)$ for the character

$$x \longmapsto \theta(w^{-1}(x)) \quad \text{of} \quad w(U).$$

We say that a subgroup H of G is <u>decomposable with respect to (M,U)</u> if

$$H \cap (MU) = (H \cap M) \cdot (H \cap U) \quad .$$

Let us make the following assumption:

(D) For any $w \in W$ each of the groups $w(P)$, $w(M)$, and $w(U)$ is decomposable with respect to (N,V) while each of $w^{-1}(Q)$, $w^{-1}(N)$, and $w^{-1}(V)$ is decomposable with respect to (M,U).

Now for any $w \in W$ we define the functor

$$\Phi_w : \mathcal{A}(M) \longrightarrow \mathcal{A}(N).$$

Consider the condition.

(*) The characters $w(\theta)$ and ψ coincide on $w(U) \cap V$.

If (*) does not hold, we put $\Phi_w = 0$. If (*) holds then Φ_w is defined as follows. Put

$$M' = M \cap w^{-1}(N), \quad N' = w(M') = w(M) \cap N,$$

$$V' = M \cap w^{-1}(V), \quad U' = N \cap w(U),$$

ψ' = restriction of $w^{-1}(\psi)$ to V', and

θ' = restriction of $w(\theta)$ to U'.

By (D), there are defined the functors

$$r_{V',\psi'} : \mathcal{A}(M) \longrightarrow \mathcal{A}(M'),$$

$w : \mathcal{A}(M') \longrightarrow \mathcal{A}(N')$ (transfer of structure by w), and $\quad i_{U',\theta'}: \mathcal{A}(N') \longrightarrow \mathcal{A}(N)$.

We set

$$\varphi_w = i_{U',\theta'} \circ w \circ r_{V',\psi'}: \mathcal{A}(M) \longrightarrow \mathcal{A}(N).$$

<u>Theorem.</u> The functor $F = r_{V,\psi} \circ i_{U,\theta} : \mathcal{A}(M) \longrightarrow \mathcal{A}(N)$ is isomorphic to the direct sum of the functors φ_w, $w \in W$.

This theorem even in a more general setting (for locally compact 0-dimensional groups) was proved in [1], § 5. The more elementary proof of the fact that F and $\sum \varphi_w$ coincide as mappings from $R(M)$ to $R(N)$ (this is sufficient for the purposes of this work) can be obtained by a computation of characters via the formulas from 8.2; we leave this to the reader.

When $U = V = \{e\}$, our Theorem is just the well-known Mackey theorem ([9] , 7.4); the conditions (D) and (*) for any $w \in W$ hold tautologically.

A3.2. Let us apply Theorem A3.I to verify the axiom (H) for the algebra $R(S)$ (see 6.2). We have to compute the composition

$$R(S)_{k'} \otimes R(S)_{l'} \xrightarrow{m} R(S_n) \xrightarrow{m^*_{k,l}} R(S)_k \otimes R(S)_l$$

(here $n = k+l=k'+l'$). By definition, this composition can be rewritten as

$$R(S_{k'} \times S_{l'}) \xrightarrow{i_{e,I}} R(S_n) \xrightarrow{r_{e,I}} R(S_k \times S_l) .$$

So we apply Theorem A3.I in the case when $G = S_n$, $M = S_{k'} \times S_{l'}$, $N = S_k \times S_l$, and $U = V = \{e\}$. According to A3.I, we must choose a set W of representatives of double cosets

$$S_k \times S_l \backslash S_n / S_{k'} \times S_{l'} .$$

Let us give the more general result. For any ordered partition

$\alpha = (a_1, \dots, a_r)$ of n let $S_\alpha \subset S_n$ be the subgroup consisting of permutations preserving blocks of α i.e. subsets

$$I_1 = \{1, \dots, a_1\},\ I_2 = \{a_1+1, \dots, a_1+a_2\}, \dots,\ I_r = \{a_1 + \dots + a_{r-1}+1, \dots, n\}.$$

Proposition. Let $\alpha = (a_1, \dots, a_r)$ and $\beta = (b_1, \dots, b_s)$ be two ordered partitions of n with blocks $I_1, \dots, I_r$ and $J_1, \dots, J_s$ respectively. Then double cosets $S_\beta \backslash S_n / S_\alpha$ are parametrized by matrices

$$K = (k_{ij})_{1 \leq i \leq r,\ 1 \leq j \leq s}$$

such that $k_{ij} \in \mathbb{Z}^+$, $\sum_j k_{ij} = a_i$ for $1 \leq i \leq r$, and $\sum_i k_{ij} = b_j$ for $1 \leq j \leq s$ (cf. 3.17 (c)). Namely, the double coset Ω_K corresponding to a matrix $K = (k_{ij})$ consists of permutations $w \in S_n$ such that

$$|w(I_i) \cap J_j| = k_{ij} \quad \text{for all } (i,j) .$$

One can choose as a representative of Ω_K the permutation w_K which acts on each block of the partition

$$(k_{11}, k_{12}, \dots, k_{1s}, k_{21}, \dots, k_{2s}, \dots, k_{r1}, \dots k_{rs})$$

by a shift, sending the block k_{ij} to the i-th place in J_j.

Proof. For any $s \in S_\alpha$, $s' \in S_\beta$ and $w \in S_n$ we have

$$s'ws(I_i) \cap J_j = s'w(I_i) \cap J_j =$$
$$= s'\left[w(I_i) \cap s'^{-1}(J_j)\right] = s'\left[w(I_i) \cap J_j\right].$$

Hence

$$|s'ws(I_i) \cap J_j| = |w(I_i) \cap J_j|,$$

i.e. the number $\left|\, w(I_i) \cap J_j \,\right|$ does not change when w varies in its double coset $S_\beta \backslash S_n / S_\alpha$. Conversely, it is easy to verify that if

$$\left|\, w(I_i) \cap J_j \,\right| = \left|\, w'(I_i) \cap J_j \,\right| \quad \text{for all } i,j$$

then w and w' lie in the same double coset. Finally, the inclusion $w_K \in \Omega_K$ is evident.

Q.E.D.

In particular, double cosets $S_k \times S_1 \backslash S_n / S_{k'} \times S_{1'}$ are parametrized by matrices

$$K = \begin{pmatrix} k_{II} & k_{I2} \\ k_{2I} & k_{22} \end{pmatrix}$$

such that

$$k_{II} + k_{I2} = k' \; , \; k_{2I} + k_{22} = 1' \; , \; k_{II} + k_{2I} = k \quad \text{and}$$
$$k_{I2} + k_{22} = 1 \; ;$$

so we choose W consisting of all w_K .

Put $w = w_K$. Then under the notation of AI.I we have

$$M' = S_{(k_{II}, k_{I2}, k_{2I}, k_{22})} \quad \text{and} \quad N' = S_{(k_{II}, k_{2I}, k_{I2}, k_{22})} \; ,$$

while the functor $w: \mathcal{A}(M') \longrightarrow \mathcal{A}(N')$ sends

$$\pi_{II} \otimes \pi_{I2} \otimes \pi_{2I} \otimes \pi_{22} \quad \text{to} \quad \pi_{II} \otimes \pi_{2I} \otimes \pi_{I2} \otimes \pi_{22}$$

(here $\pi_{ij} \in \mathcal{A}(s_{k_{ij}})$). Therefore

$$\Phi_w(\pi \otimes \pi') = m^*_{k_{II}, k_{I2}}(\pi) \cdot m^*_{k_{2I}, k_{22}}(\pi')$$
$$(\pi \in \mathcal{A}(S_{k'}), \; \pi' \in \mathcal{A}(S_{\ell'})).$$

Adding these expressions together for all K and then for all (k,l) with k+l=n, we see that

$$m^*(\pi\cdot\pi') = m^*(\pi)\, m^*(\pi'),$$

as desired.

A3.3. The axiom (H) for the algebra $R(S[G])$ (see 7.2) can be verified in a quite similar way as in A3.2. It suffices to observe that

$$S_n[G] = S_n\cdot G^n \quad\text{and}\quad S_k[G]\times S_l[G] = S_{(k,l)}\cdot G^n$$

(see 7.I), so

$$S_k[G]\times S_l[G]\backslash S_n[G]/S_{k'}[G]\times S_{l'}[G] = S_{(k,l)}\backslash S_n/S_{(k',l')}$$

and one can choose the same W as in A3.2.

A3.4. Now we prove that the restriction

$$R : R(S[G]) \longrightarrow R(S) \qquad \text{(see 7.I0)}$$

is a ring homomorphism. If $\pi\in R(S_k[G])$, $\rho\in R(S_l[G])$, and k+l=n, them $R(\pi\cdot\rho)$ is by definition the image of $\pi\otimes\rho\in R(S_{(k,l)}[G])$ under the action of the composition

$$R(S_{(k,l)}[G]) \xrightarrow{i_{e,I}} R(S_n[G]) \xrightarrow{r_{e,I}} R(S_n)$$

This composition can be computed by means of Theorem A3.I. We have $W=\{e\}$ so our composition equals

$$R(S_{(k,l)}[G] \xrightarrow{r_{e,I}} R(S_{(k,l)}) \xrightarrow{i_{e,I}} R(S_n),$$

i.e. it sends $\pi\otimes\rho$ to $R(\pi)\cdot R(\rho)$, as desired.

A3.5. Now we verify the axiom (H) for the algebra $R(q)$ (see 9.I). As in A3.2, the composition

$$R(q)_{k'} \otimes R(q)_{l'} \xrightarrow{m} R(q)_n \xrightarrow{m^*_{k,l}} R(q)_k \otimes R(q)_l$$

can be rewritten as

$$R(G_{k'} \times G_{l'}) \xrightarrow{i_{U_{k',l'},I}} R(G_n) \xrightarrow{r_{U_{k,l},I}} R(G_k \times G_l)$$

(see 9.I). So we apply Theorem A3.I in the next situation: $G=G_n$, $M = G_{k'} \times G_{l'}$, $U = U_{k',l'}$, $\theta = I$, $N = G_k \times G_l$, $V = U_{k,l}$, and $\psi = I$. We have to choose a set W of representatives of double cosets

$$(G_k \times G_l)\, U_{k,l} \backslash G_n / (G_{k'} \times G_{l'})\, U_{k',l'}$$

As in A3.2, let us give the more general result. For any ordered partition $\alpha = (a_I,\dots,a_r)$ of n denote by P_α the subgroup of G_n consisting of all cellular upper triangular matrices with cells of lengths $a_I,\dots,a_r$ (from up to down).

<u>Proposition</u> (Bruhat Decomposition). Let α and β be two ordered partitions of n. The natural embedding $S_n \hookrightarrow G_n$ ($w \longmapsto (\delta_{i,w(j)})$) induces the bijection

$$S_\beta \backslash S_n / S_\alpha \xrightarrow{\sim} P_\beta \backslash G_n / P_\alpha$$

Thus, the representatives of double cosets $P_\beta \backslash G_n / P_\alpha$ can be chosen in accordance to Prop. A3.2.

For the proof see [29] , ch. IV, § 2, items 2 and 3.

Q.E.D.

The further arguments are similar to those in A3.2. Note that A3.I (*) holds for any $w \in W$, since each of θ and ψ equals I; the condition (D) for any $w \in W$ can be verified directly.

A3.6. Let us verify that the form $\delta : R(q) \longrightarrow \mathbb{Z}$ (see 9.4) is multiplicative. Let $\pi \in R(G_k)$, $\rho \in R(G_l)$, and $k+l=n$. By definition (see 9.I, 9.4), $\delta(\pi\cdot\rho)$ is obtained by applying the composition

$$R(G_k x G_l) \xrightarrow{i_{U_{k,l}, I}} R(G_n) \xrightarrow{r_{U_n, \psi}} R(G_o) = \mathbb{Z}$$

to $\pi \otimes \rho \in R(G_k x G_l)$. To compute this composition we apply Theorem A3.I in the next situation;

$$G = G_n, \ M = G_k x G_l, \ U = U_{k,l}, \ \theta = I, \ N = G_o = \{e\},$$

$V = U_n$, and ψ is defined in 9.4.

We have to choose a set W of representatives of double cosets

$$U_n \backslash G_n / P_{(k,l)} \qquad \text{(see A3.5).}$$

Let $B = B_n$ be the subgroup of all upper triangular matrices in G_n, and $D = D_n$ be the diagonal subgroup in G_n. We have $B = P_{(I^n)}$ (see A3.5), so the Bruhat decomposition implies that

$$B \backslash G_n / P_{k,l)} = S_n / S_{(k,l)}$$

On the other hand, $B = UD$, and D is normalized by any permutation matrix, hence

$$U \backslash G_n / P_{(k,l)} = B \backslash G_n / P_{(k,l)}$$

Thus, we can choose W as the set of representatives of $S/S_{(k,l)}$; we choose it in accordance to Prop. A3.2. It is easy to see that W consists of permutations w such that

$$w(1)<w(2)<\ldots<w(k) \quad \text{and} \quad w(k+1)<w(k+2)<\ldots<w(n).$$

The condition (D) from A3.1 for all these w can be verified directly.

Now consider the condition (*) from A3.1. Clearly, $w(U)$ consists of matrices (u_{ij}) such that $u_{ii}=1$, and for $i\neq j$ the entry u_{ij} can be non-zero only if $w^{-1}(i)\leq k<w^{-1}(j)$. It follows that if $w^{-1}(i)\leq k<w^{-1}(i+1)$ for some i then $\psi\neq 1$ on $w(U)\cap V$ i.e. (*) does not hold. Thus, (*) holds only if

$$w(\{1,\ldots,k\})=\{l+1,\ldots,n\} \quad \text{and} \quad w(\{k+1,\ldots,n\})=$$
$$=\{1,\ldots,l\}$$

There exists the unique such $w\in W$:

$$w(i)=\begin{cases} l+i & \text{if } i\leq k \\ i-k & \text{if } i>k \end{cases}$$

The corresponding subgroup $V'=M\cap w^{-1}(V)$ (see A1.1) equals $U_k\times U_l\subset G_k\times G_l=M$, while $\psi'=\psi$. By Theorem A1.1 and 8.1 (c),(d):

$$\delta(\pi\cdot\rho)=\varepsilon(\pi)\cdot\varepsilon(\rho),$$

as desired.

A3.7. In conclusion we compute the induction of representations of G_{n-1} to P_n (see 13.3). We have to compute the compositions

$$r_{V,1}\circ i_{e,1} : \mathcal{A}(G_{n-1})\longrightarrow\mathcal{A}(G_{n-1}) \quad \text{and}$$

$$r_{V,\psi}\circ i_{e,1} : \mathcal{A}(G_{n-1})\longrightarrow\mathcal{A}(P_{n-1}).$$

Let us apply Theorem A3.I in the next two situations:

I. $G = P_n, M = G_{n-I}$, $U = \{e\}$, $N = G_{n-I}$, $V = U_{n-I,I}$ and $\psi = I$.

II. $G = P_n, M = G_{n-I}$, $U = \{e\}$, $N = P_{n-I}$, $V = U_{n-I,I}$, and ψ is defined in 9.4.

In each case $W = \{e\}$, and (D) and (*) from A3.I hold automatically. Theorem A3.I implies, that the first composition is the identity functor, while the second one equals $r_{V,\psi}$. This completes the proof of Prop. I3.3.

References

I. I.N.Bernstein, A.V.Zelevinsky. Induced representations of reductive $\mathfrak{p}$-adic groups I. Ann. Sc. Éc. Norm. Sup., t. IO, no. 4, 44I-472, I977.

2. A.V.Zelevinsky. Induced representations of reductive $\mathfrak{p}$-adic groups II. On irreducible representations of GL(n). Ann. Sc. Éc. Norm. Sup., t. I3, no. 2, 165-210,1980.

3. A.V.Zelevinsky. On representations of the general linear and affine groups over a finite field. Uspekhi Mat. Nauk, 32, no. 3, I59-I60, I977.

4. J.A.Green. The characters of the finite general linear groups. Trans. Amer. Math. Soc., 80, no. 2, 402-447, I955.

5. L.Geissinger. Hopf algebras of symmetric functions and class functions, in "Combinatoire et représentation du groupe symétrique", Lecture Notes in Math., 579, I68-I8I, Springer-Verlag, I977.

6. A.V.Zelevinsky. The representation ring of the groups GL(n) over a $\mathfrak{p}$-adic field. Funkc. Anal., II, no. 3, 78-79, I977.

7. D.K.Faddeev. The complex representations of the general linear group over a finite field. Zapiski Nauchn. Semin. LOMI, 46, 64-88, I974.

8. J.W.Milnor, J.C.Moore. On the structure of Hopf algebras. Annals. of Math., 81, no. 2, 2II-264, I965.

9. J.-P.Serre.Représentations linéaires des groupes finis. Hermann, Paris, I967.

IO. D.Knutson. λ-rings and the representation theory of the symmetric group. Lecture Notes in Math., 308, Springer-Verlag, I972.

II. J.S.Frame, G. de B.Robinson, R.M.Thrall. The hook graphs of the symmetric group. Canad, J. Math., 6, 3I6-324, I954.

I2. A.Kerber. Representations of permutation groups I. Lecture Notes in Math., 240, Springer-Verlag, I97I.

I3. I.M.Gelfand, M.I.Graev. The construction of irreducible representations of simple algebraic groups over a finite field, Dokl. Akad. Nauk U.S.S.R., I47, no. 3, 529-532, I962.

I4. S.I.Gelfand. Representations of the general linear group over a finite field. Mat. Sbornik, 83, no. I, I5-4I,I970.

I5. T.A.Springer. Characters of special groups, in "Seminar on algebraic groups and related finite groups", Lecture Notes in Math., 131, Springer-Verlag, I970.

I6. Z.Ohmori. On the Schur indices of GL(n,q) and SL(2n+I,q). J.Math. Soc. Japan, 29, no.3, 693-707, I977.

I7. D.K.Faddeev. On the complex representations of the general affine group over a finite field. Dokl. Akad. Nauk U.S.S.R., 230, no. 2, 295-297, I976.

I8. E.Thoma. Die Einschränkung der Charactere von GL(n,q) auf GL(n-I,q). Math. Z., II9, 32I-338, I97I.

I9. G.D.James, M.H.Peel. Specht series for skew representations of symmetric groups. J. Alg., 56, 343-364, I979.

20. A.V.Zelevinsky. A generalization of the Littlewood-Richardson rule and the Robinson-Schensted-Knuth correspondence. To appear in J. of Algebra.

2I. R.A.Liebler, M.R.Vitale. Ordering the partition character of the symmetric group. J. Alg., 25, no.3, 487-489, I973.

22. D.E.Littlewood. The theory of group characters and matrix representations of groups, 2-nd ed. Oxford, Clarendon, I950.

23. R.Stanley. Theory and applications of plane partitions. Part I. Studies in Applied Math., 50, I67-I88, I97I.

24. T.A.Springer. Cusp forms for finite groups, in "Seminar on algebraic groups and related finite groups", Lecture Notes in Math., I3I, Springer-Verlag, I970.

25. R.Steinberg. A geometric approach to the representations of the full linear group over a Galois field. Trans. Amer. Math. Soc., 7I, no. 2, 274-282, I95I.

26. C.W.Curtis. Truncation and duality in the character ring of a finite group of Lie type. J. Alg. 62, 320-332 (1980).

27. D.Alvis. The duality operation in the character ring of a finite Chevalley group. Bull. Amer. Math. Soc,, New Series, I, 907-9II (1979).

28. R.Gow. Schur indices of some groups of Lie Type. J. Alg., 42, No. I, I02-I20, I976.

29. N.Bourbaki. Groupes et algèbres de Lie, Chaps. IV,V, VI, Hermann, Paris, I968.

30. D.Knuth. The art of computer programming, v. 3, Sorting and Searching. Addison-Wesley, I973.

<u>INDEX OF NOTATION</u>

INDEX